Paleoceanography of the Mesozoic Alpine Tethys

KENNETH J. HSÜ
Geological Institute
Swiss Federal Institute of Technology
Zurich, Switzerland

THE GEOLOGICAL SOCIETY OF AMERICA

SPECIAL PAPER 170

The Special Paper series was originally made possible through the bequest of Richard Alexander Fullerton Penrose, Jr.

Library of Congress Catalog Card Number 75-32124
I.S.B.N. 0-8137-2170-9

Published by
THE GEOLOGICAL SOCIETY OF AMERICA, INC.
3300 Penrose Place
Boulder, Colorado 80301

Printed in The United States of America

Contents

Abstract

Three important Alpine pelagic lithologic types are ammonitico rosso, Radiolarite, and Maiolica, which range in age from late Early Jurassic to Early Cretaceous.

Ammonitico rosso is in many places red nodular limestone, lying above continental crust. This pelagic deposit bears some resemblance to some Holocene nodular deposits in the eastern Mediterranean. A somewhat restricted Middle Jurassic Alpine Tethys sea has been postulated to account for the occurrence of the "home-made" bottom currents that led to partial dissolution of aragonite and precipitation of calcite nodules.

The sudden appearance of the radiolarite facies in Late Jurassic time may signify a major tectonic event that sufficiently altered the Tethyan paleogeography to permit the intrusion of an equatorial current into the Alpine realm. The radiolarite facies is commonly believed to have been deposited beneath the calcite compensation depth, but calcitic aptychi have been found in radiolarite, and Aptychus Limestone is intimately associated with this mainly siliceous formation. The depths of the Radiolarite sea during Late Jurassic time were probably closer to the lysocline for aragonite. Because the Radiolarite is the first sediment deposited on ophiolite, one might postulate that the Tethys at the time of Radiolarite deposition had an oceanic depth comparable to that in regions of newly formed crust. Both lines of reasoning have led me to suggest that the Tethys in Late Jurassic time had a depth of 2,500 $\pm$ 200 m, not 4,000 to 5,500 m as postulated previously by some authors.

The Maiolica Formation is pelagic limestone with slump interbeds and intraformational chert nodules. The radiolarite-limestone succession in the Alpine Tethys is exactly the opposite of the simple model constructed by plate stratigraphers, who usually found a limestone-radiolarite-red clay succession on an aging crust. It is possible that excessive blooms of siliceous faunas may have created a silica crisis at the end of Jurassic time. The Alpine Tethys, being disadvantageously situated at the end of the Jurassic equatorial current system, could no longer be given enough silica to produce radiolarians. In their place, a calcareous flora of nannofossils flourished, resulting in the change from Radiolarite sedimentation to Maiolica sedimentation.

Considering the rate of the silica transport, chert in the Maiolica deposit could not have been formed by submarine diagenesis. Nodules took millions of years to grow in formations in which concentration gradients existed between the domain where amorphous biogenic silica was being dissolved and the nuclei around which quartz or cristobalite was being precipitated. *Key words: ammonitico rosso, Radiolarite, Maiolica, paleobathymetry, nodular limestone, compensation depth, lysocline, chertification.*

A Tale of Two *Challengers*

In his work on the origin of the Alps, Suess (1875) criticized the Hallian concept of neritic sedimentation in geosynclines. Suess believed that the Alpine sequence was remarkable not because it was thicker than the coeval platform deposits on the Bohemian massif but because the geosynclinal sediments were typified by their pelagic character. Following a simplistic scheme suggested in 1845 by the French geologist Prevost that gravel, sandstone, clay, and limestone represented deposits of increasing water depths, Suess cited the thick Triassic carbonate strata as evidence of pelagic sedimentation in the Alpine geosyncline. He was wrong in part, as some of those strata were sediments in shelf seas and on tidal flats (see Fischer, 1963). Also wrong was his speculation that the Mesozoic benthonic faunas of the Alps were relic forms that had survived in oceanic deeps. Nevertheless, Suess managed to reap a grain of truth, for pelagic rocks do exist in the Alps.

Suess's young associate Melchior Neumayr waited until the return of the H.M.S. *Challenger* from her historic voyage to make his point. Neumayr (1886) compared Upper Jurassic radiolarian chert of the Alps to Holocene radiolarian ooze sampled by the *Challenger* from the equatorial Pacific and Indian Oceans. This interpretation stood the test of time, and Neumayr's belief that the Tertiary radiolarite of Barbados (Eocene Oceanic Formation) had been uplifted from a deep ocean bottom could be considered proven by results of deep-sea drilling by the *Glomar Challenger* (Bader and others, 1970).

Associated with Alpine radiolarite are aptychus beds. The aptychi are odd calcitic fossils, which bear some superficial resemblance to bivalves (Fig. 1). As early as 1829, Rüppell correctly identified them as the opercular structures of ammonoids (see review by Renz, 1972). The most remarkable aspect of aptychi occurrences is, however, that operculars in an aptychus bed are far more numerous than phragmocones, which are the aragonitic chambered shells of ammonites. What happened to the phragmocones? Theodor Fuchs, a student of Suess, provided an answer to this puzzle in two brilliant essays written in 1877 and 1882, shortly after the return of the H.M.S. *Challenger*. Fuchs was aware of the work by Rose (1858) on the difference in solubility between calcite and aragonite. When Fuchs learned in 1877 of the findings by the *Challenger* on $CaCO_3$ dissolution in deep seas, he immediately concluded that the preferential preservation of the calcitic operculars in an aptychus bed resulted from a selective dissolution of aragonitic phragmocones. Learning that the *Challenger* had sampled pteropods with delicate aragonitic shells from the Atlantic at depths of as much as 4,000 m, Fuchs (1882, p. 510) could not help but postulate a very great deep as the depositional environment of the Aptychus Limestone; its pelagic assemblage includes aptychi, belemnites, crustaceans, fish, and other fossils with calcitic or phosphatic tests, and it is characterized by the almost total absence of aragonitic shells of any kind.

Gustav Steinmann (1905) was the first to recognize the common association of the Alpine pelagic sediments with ophiolite. In his 1905 work he reported having noted in 1889 the association of radiolarite with ophiolite in the area of Iberg Klippes in the Prealps. It was then popular in some circles to dismiss the significance of this association, as radiolarite or "hornstone" had been considered a product of contact metamorphism. Steinmann was surprised to see again the association of radiolarite and pelagic limestone associated with serpentine in the Apennines near Florence. After the 1891 International Geological Congress, Steinmann took a field trip to San Francisco. By then he was certain that the association of ophiolite and deep-sea sediments was no coincidence. He cited many occurrences of this association in all parts of the world

Figure 1. Aptychus—calcitic opercular of an ammonite. Specimen shown is attached to an ammonite. It is not uncommon, however, that only aptychus is preserved in Alpine pelagic rock. Occurrences of aptychus without chambered shell bear witness to preferential dissolution of aragonite fossils in Jurassic Alpine Tethys. (Photo courtesy of R. Gygi.)

in his 1905 classic work; he proposed then that ophiolite was a suite of eruptive rocks typical of deep-sea environments.

Suess, as well as John Murray and Emil Haug, was mentioned by Steinmann in 1905 as being skeptical of the deep-sea origin of carbonate-free radiolarites.[1] Yet in the third volume of *Das Antlitz der Erde*, published in 1909, Suess fully agreed with Steinmann that the Alpine ophiolite was emplaced in a deep-sea environment. Suess compared the Alpine ophiolite to submarine basalt from the Atlantic, to gabbro in Iceland, and to peridotite on St. Paul Rock (1909, p. 645). Noting further that the carbonate base of shark teeth in the Aptychus Limestone has been dissolved, he shared Fuch's view that the Late Jurassic–Early Cretaceous Tethys must have been 4,000 m deep.

The Austrian government sent its vessel *Pola* on four cruises to the eastern Mediterranean Sea in 1890–1893. The prevailing opinion in Austria at the time, however, seemed to be that oceanographic researches were only for physicists and zoologists, and that geologists, whose domain was on land, would be as misplaced on an oceanographical vessel as a cavalry officer (Tietz, 1897). The *Pola* expeditions found large areas of the Mediterranean sea floor that were covered by a hard crust. A chemist noted that the crust was rich in iron and manganese and that its genesis was related to dissolution and precipitation induced by bottom currents (Natterer, 1899). Some dredged samples were brought to Fuchs; he described some cylindrical objects but failed to perceive their significance. Fuchs did recognize, however, the need for geologists to acquire a berth on an oceanographical vessel; he wrote (1898, p. 108)

> It is true that the object of study of geologists is the continents and not the oceans, but it is just as true that a very large part of the continents was formed of sediments laid down in oceans. It would be impossible to interpret the ancient rocks without an exact knowledge of the depositional environments and genesis of the Recent sediments on modern-day sea-floor.

Fuchs was pamphleteering in support of a proposal to initiate an "international floating laboratory" for marine geology. The project was formulated by Nikolai Andrussov, John Murray, Karl v. Zittel, Johannes Walther, François Forel, and Ernst Haeckel, some of the most prominent scientists then, and the proposal was unanimously adopted by the members present at the International Geological Congress in St. Petersburg in 1897. Unfortunately, these pioneers were way ahead of their time, and none could have lived long enough to see the *Glomar Challenger* and the International Program for Ocean Drilling.

Suess died in 1914, and the Hallian concept of geosynclines prevailed. Pelagic rocks are indeed present on land, but did not Jeffreys (1877) prove that

---

[1]It was probably a misunderstanding on Steinmann's part. Hugh Jenkyns (1974, personal commun.) discovered on the reprint of Steinmann's paper in the Sedgwick Museum, Cambridge, "an alteration in Steinmann's own handwriting [changing the name] from Suess to Geikie."

the British Chalk was a shallow-water pelagic deposit? Did not Walther (1897) point out that the Alpine radiolarite was not oceanic sediment? Walther's conclusion was strengthened by an assessment by John Murray, of the *Challenger*. A. W. Grabau (Grabau and O'Connell, 1917) also concluded that pelagic sedimentary strata were not necessarily deposited in deep waters. Black shale, the Appalachian equivalent of the Alpine pelagic sedimentary rocks, was interpreted variously as shallow to moderately deep but never as abyssal (Ulrich, 1911; Twenhofel, 1939; Ruedemann, 1934).

Steinmann did not lose sight of the fact that the bathymetric significance of Alpine pelagic rocks is a critical piece of evidence for the solution of the controversy on the permanency of continents and oceans. He continued to argue for the oceanic character of the Alpine pelagic rocks (Steinmann, 1925, 1927). He did not have many followers, even in Europe, because in the 1930s and 1940s, as Trümpy commented later (1960, p. 866), "European geologists [also] tended to consider practically all visible sedimentary rocks as having been formed in shallow water and the radiolarites suffered the same 'fate.' " Perhaps the major reason that led geologists to reject the deep-sea interpretation of eugeosynclinal radiolarite was the association of radiolarite with coarse-grained clastic beds (see, for example, Davis, 1918; Grunau, 1947). Steinmann did not succeed in dismissing the problem when he misidentified slump beds as tectonic breccia (1925, p. 452–456). The puzzle was solved only after the introduction of the concept of deep-sea turbidite sedimentation. By 1960, there was no reason left "to reject Steinmann's views on the deep-sea origin of radiolarian cherts in the Alps" (Trümpy, 1960, p. 866).

The Alpine pelagic rocks range in age mainly from Early Jurassic to Early Cretaceous. The ophiolite masses are, however, difficult to date. They occur mainly as blocks in mélange, and their relation to the pelagic sediments was not easy to interpret. The mélange nature of the ophiolite-bearing terranes was very confusing to pioneers on both sides of the Atlantic. Misled by the erroneous age assignments of ophiolite, Steinmann did not conclude that the ophiolite masses were fragments of oceanic floor mixed with overlying oceanic sediment. Instead, he thought that the Alpine ophiolite was intrusive and extrusive into an older, pre-existing Alpine pelagic rock (1905, 1927). Steinmann (1913) was puzzled by the absence of metamorphism where the contact between radiolarite and ophiolite was observed. In California, Franciscan ophiolite was similarly interpreted as intrusive, and glaucophane schist was thought to be a contact-metamorphic product of such intrusions (see, for example, Taliaferro, 1943). Several elaborate schemes of geosynclinal magmatism were formulated (Kossmat, 1921; Scheumann, 1932; reviewed by Knopf, 1948). In fact, even in the early 1960s, the Alpine ophiolite rocks were considered intrusive into, synchronous with, or younger than the Alpine radiolarite (Trümpy, 1960; Aubouin, 1964), and the earliest proponents of plate tectonics still envisioned ophiolite as forming by synchronogenic intrusions (Dewey, 1969). The current concept that ophiolite represents oceanic crust and mantle has been developed in the past decade through the combined efforts of marine geophysical, petrological, structural, and deep-sea drilling workers (see review by Moores, 1973).

The renaissance of the Alpine pelagic sedimentology coincided more or less with the revival of the idea of continental drift in the 1960s. Trümpy's essay in 1960 was followed by a review by Garrison and Fischer in 1969. Finally in 1970 the *Glomar Challenger* cruised the Atlantic and sampled pelagic sediment and oceanic basement rock that are remarkably similar to the Alpine pelagic and ophiolitic sequence (see, for example, Hollister, Ewing, and others, 1972; Ryan, Hsü, and others, 1973). Bernoulli's (1972) comparative study of the Alpine and Atlantic Mesozoic facies dispelled the last lingering doubt that the Tethys was oceanic. We might at this stage be satisfied and conclude that the Alpine pelagic sediments were deposited in a Mesozoic spreading ocean. We might be tempted to use some simple rules of plate stratigraphy to reconstruct the paleobathymetry of the Alpine geosyncline (Fig. 2). Unfortunately, nowhere do we find a simple ophiolite–chalk–radiolarite–red clay sequence in the Alpine rocks, as predicted by a model of sedimentation on spreading ridges. Many sedimentological problems about this area

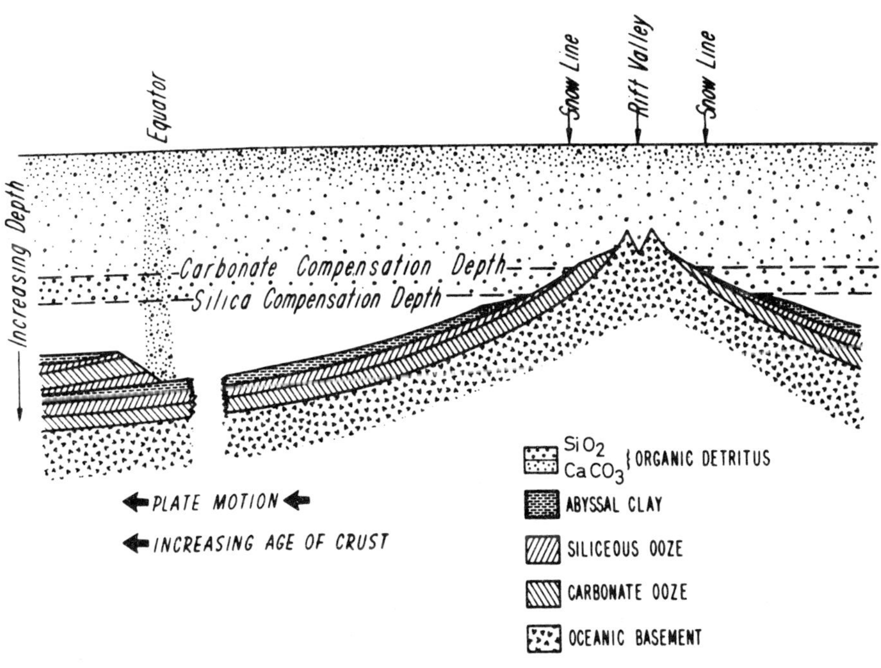

Figure 2. Idealized model of oceanic pelagic sedimentation (after Heezen and others, 1973). This model assumes presence of silica compensation depth beneath carbonate compensation depth. Normal facies succession above oceanic basement should be carbonate ooze–siliceous ooze–red clay. Facies changes occurred at times when spreading crust sank below calcite and silica compensation depths. Stratigraphic succession in Alps cannot be explained by such an idealized model.

were discussed during a symposium on pelagic sediments in Zurich in 1973 (see Hsü and Jenkyns, 1974). This paper reviews these problems, and I speculate here on three particularly puzzling questions concerning the Alpine pelagic sediments: the origin of nodular limestone, the depth of radiolarite sedimentation, and the cause of penecontemporaneous chertification.

Origin of Nodular Limestone

It is well established that the Triassic Tethys sea did not find its way into western Europe. The Alpine-Apennine geosyncline was mainly a Jurassic creation (Trümpy, 1960), and the birth of this mediterranean sea was apparently induced by the eastward motion of Africa relative to Europe in connection with the opening of the central Atlantic (Smith, 1971; Hsü, 1971; Dewey and others, 1973). The Jurassic Tethys can be divided into three zones (Bernoulli, 1972): (1) the northern continental margin, represented by the Helvetic and Ultrahelvetic zones of the Alps, (2) the central oceanic trough, represented by the ophiolite zones, and (3) the southern Tethys margin, fringing the several microcontinents between the European and African plates. The southern Tethys margin is represented by the external zones of the Apennines, the Dinarides and Hellenides, the southern Alps, the Austroalpine thrust sheets, and the internal Carpathians.

The first widespread pelagic sediment of the Jurassic Tethys is nodular limestone, which is locally given names such as ammonitico rosso, Knollenkalk, Adnetschichten, and cephalopod limestone (see Fig. 3). These pelagic facies are commonly underlain by crinoidal limestone, marking the transition from shallow-water to deep-water sedimentation (Garrison and Fischer, 1969; Schlager, 1974). Aside from their pelagic nature and rich ammonite fauna, these formations have two interesting features: (1) their bedding surfaces are uneven and give a nodular appearance (Fig. 4), and (2) in many places they are reddish brown.

The red color of the ammonitico rosso is obviously related to its hematite content. Sediment of the same facies but younger has been encountered by JOIDES drilling in the western North Atlantic (Bernoulli, 1972). It is also red because it is hematitic, and the iron in the earliest Atlantic sediment has been thought to have an origin similar to that of the amorphous iron oxides found in the sediment on active ocean ridges (Lancelot and others, 1972).

Boström (1967) drew attention to the presence of metal-rich sediments at the crest of the East Pacific Rise. Later Boström and others (1969) reported occurrences of them in the Indian Ocean as well. They further noted that aluminum-poor ferromanganoan sediments are more common on the faster spreading ridges of the Pacific and Indian Oceans than on the Mid-Atlantic Ridge. Deep-sea–drilling cruises confirmed the existence of a basal ferruginous facies, lying directly above the basalt basement in various parts of the Atlantic

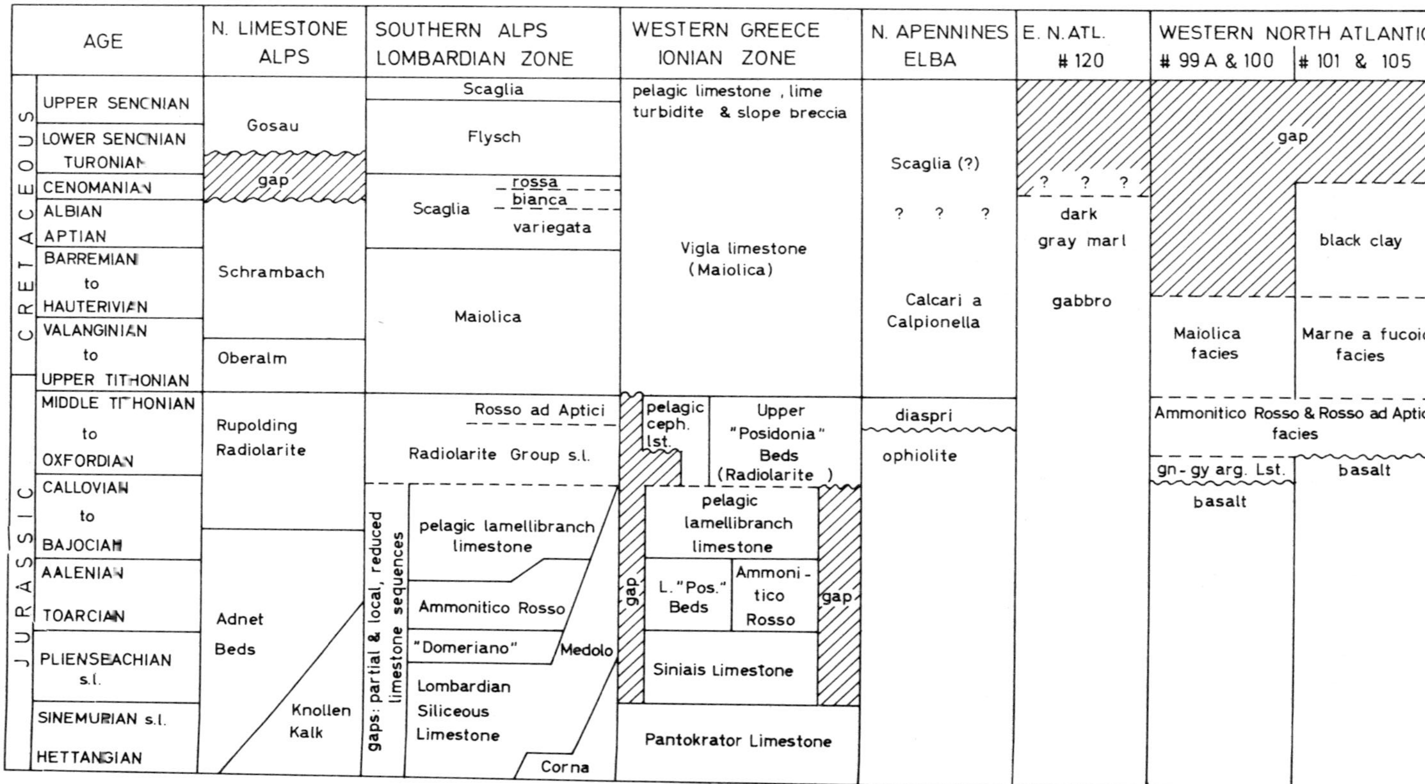

Figure 3. Simplified stratigraphic scheme of some Jurassic-Cretaceous sequences in the Mediterranean region and their facies equivalents in the North Atlantic basin (modified after Bernoulli, 1972).

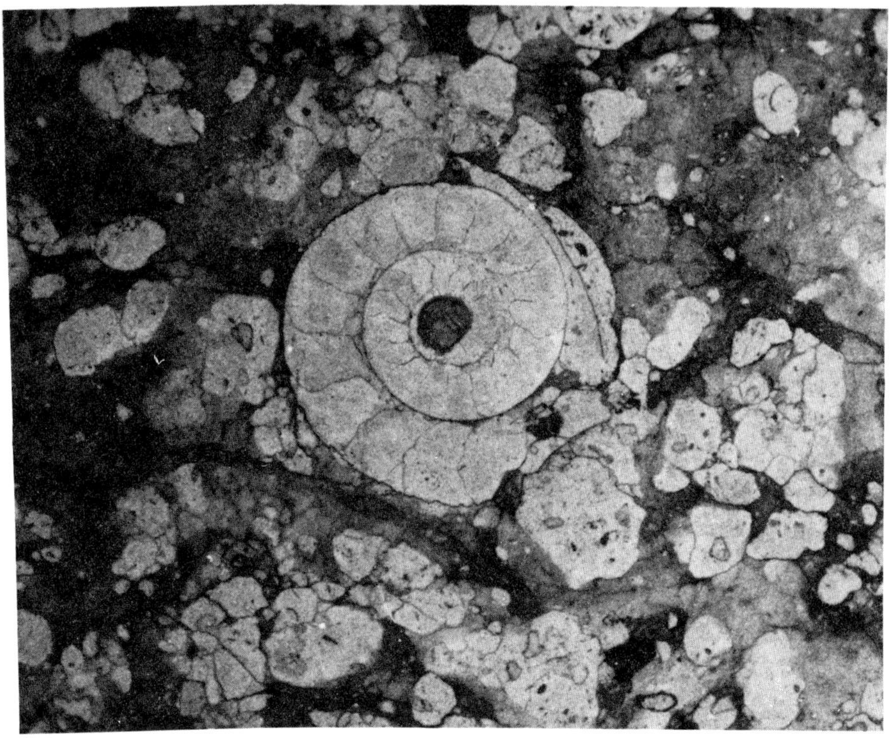

Figure 4. Ammonitico rosso. Nodules are light colored and contain more $CaC0_3$ than does matrix. Nodules may have formed during early diagenetic cementation. In contrast, stylolites (dark wavy lines) are evidence of dissolution. (Photo courtesy of F. Barbieri.)

and Pacific (Peterson and others, 1970; Maxwell and others, 1970; von der Borch and Rex, 1970; von der Borch and others, 1971; Cronan, 1973).

The Atlantic ammonitico rosso facies in JOIDES hole 105 directly overlies the basalt basement and may be comparable to active-ridge sediments. The same interpretation, however, cannot be readily applied to similar Alpine rocks, for they are present mainly in the southern Tethys realm, and they are not the first sediment directly overlying the Alpine ophiolite (Fig. 3). One might argue that hydrothermal solutions had nevertheless invaded the fragmented and foundered continental margins, even though there was not yet a mid-oceanic ridge in the embryonic Tethys. Yet we might wonder if there is a better explanation for the red color

It was noted during the JOIDES III drilling that two types of red sediment occur in the Atlantic: the basal ferruginous formation, ranging in age from Eocene to early Miocene and lying directly on basalt ("Grampus Ooze"; Hsü and Andrews, 1970), red zeolitic clay and deep-reddish-brown marl ooze ("Discovery Formation")—middle Miocene sediments apparently deposited at a very slow rate below the carbonate lysocline. With much of their $CaCO_3$ removed by dissolution, the residual marly or clayey sediments were enriched in iron. Boström (1973) also called attention to the fact that there are two types of ferromanganoan sediment: the active-ridge sediment, which owes its metal content to the degassing of magmas under a spreading ridge, and the inactive-ridge sediment, which is found on seamounts and which formed very slowly and showed much lower Fe/Ti and Fe/Al ratios than does the active-ridge sediment. Boström also noted that much inactive-ridge sediment is concretionary or nodular.

The currently popular hypothesis on the origin of nodular limestone was proposed by Hollman (1962, 1964). On the basis of evidence of dissolution (see, for example, Wähner, 1886; Steinmann, 1905), Hollman considered the uneven bedding plane of the Knollenkalk a fossilized surface of submarine dissolution or subsolution on sea floor. The numerous faunal breaks within this formation (Neumayr, 1873) were cited as further evidence of the extensive subsolution in the Jurassic Tethys (Hollman, 1962, p. 177). Because truncated ammonite shells are often coated by ferruginous crusts, it was thought that the episodes of subsolution water accompanied by a cementation of the underlying sediment formed a consolidated crust or "hardground" (Hollman, 1964; Fischer and Garrison, 1967). The importance of submarine dissolution processes has been confirmed by deep-sea drilling; in the South Atlantic, for example, the whole middle Miocene period in JOIDES hole 14 was represented by unfossiliferous hololytic red clays (Hsü and Andrews, 1970). The existence of present-day hardgrounds was also proven by the numerous dredgings from oceanographical work (Fischer and Garrison, 1967).

It should be remembered, however, that the role of submarine diagenesis below the sediment-water interface should not be neglected as a possible mechanism of dissolution. Gervitz and Friedman (1966) reported diagenetic dissolution of pteropod shells in buried pelagic deposits of the Red Sea at water depths of 816 to 1,737 m; pteropod shells are present at the sediment-water interface, but their abundance decreases with depth of burial, and only sediment-filled casts remain at a few metres subbottom. Schlager (1974) noted the presence of ammonite casts in cephalopod limestones of the Adnet or ammonitico rosso types. The casts were filled by a cement or by an internal sediment of crystal silt, closely resembling the "vadose silt" described by Dunham (1969) for subaerially altered limestones. Schlager (1974), therefore, concluded that much of the aragonite dissolution must have taken place after burial under a crust of lithified and permeable sediments comparable to hardgrounds.

Hollman and Fischer and Garrison emphasized the dissolution aspect of the nodular formation: the more marly ammonitico rosso was interpreted as having been subject to more dissolution. On the other hand, Jenkyns (1974) quoting Aubouin (1964), considered the clay content in ammonitico rosso a primary characteristic of sedimentation; marly sediments were laid down in depressions, whereas more calcareous facies covered the ridges. Jenkyns (1974) emphasized the diagenetic redistribution of carbonates, instead of a gradually decreasing carbonate content by dissolution; more soluble aragonite shells were dissolved, leaving ammonite casts behind, and ions moved away for a short distance to cement the nearby nodules. The common occurrence of ammonite molds or casts was given as evidence of aragonitic dissolution after burial, although Jenkyns admitted a certain degree of subsolution, as shown by the partial corrosion of some fossils. Similar hypotheses of diagenetic segregation had been postulated by Lucas (1966) and by Hallam (1967).

Jenkyns favored the hypothesis of diagenetic dissolution because he questioned whether the Tethys at the time of ammonitico rosso deposition was sufficiently deep to permit sea-floor dissolution of aragonite. He was particularly impressed by some stromatolitelike structures, which were thought to be algal in origin (Jenkyns, 1971). However, as Jenkyns admitted (1971, p. 339), "no algal filaments have been found." The "stromatolitic" sediments belong to a condensed sequence similar to modern hardgrounds, and the "stromatolitic" structure could well be compared to ferromanganoan encrustations. We have no positive evidence that the ammonitico rosso was deposited in photic zones. Hollman was probably correct in postulating subsolution in a moderately deep sea, but Jenkyns was probably also correct to emphasize the role of diagenetic dissolution. Significantly, both schools postulated dissolution and cementation, noted the slow rate of sedimentation, and compared the nodular bedding surface to hardgrounds on modern sea floors. There seems to be a large area of agreement on the genesis of nodules.

In 1967 Fischer and Garrison recalled the finding by the *Pola* expeditions of extensive submarine lithifications of the eastern Mediterranean sea floor; the lithified crust was dated, perhaps erroneously, as late Miocene. Fischer and Garrison referred to such lithified crust as "present-day 'hardgrounds'" and commented (1967, p. 29) that

> some of these occurrences are known to be superficial crusts overlying essentially unconsolidated sediment, and most or all of them are associated with manganese oxide crusts, nodules, or coatings. It appears that subaqueous crusts form where sedimentation is essentially at a standstill for long periods of time, and is due to long continued solution and reprecipitation processes within the chemical gradient that exists near the sediment-water interface. The Adnet beds represent a setting of this type, in which periods of more rapid lime-mud deposition alternated with periods of sea-floor solution, lithification and iron manganese enrichment.

Milliman and Müller (1973) studied in detail the submarine lithification of the eastern Mediterranean sea floor. They reported the occurrence, in addition to lithified crust, of calcareous nodules in many cores, the "cylindrical objects" studied by Fuchs (1894). The nodules range from 1 cm to more than 5 cm in diameter and form distinct nodular layers in a few cores. Milliman and

Müller established that the lithification was caused by the precipitation of magnesian calcite, and that the magnesian calcite is late Pleistocene and Holocene in age. This magnesian calcite was believed to have been precipitated from sea water at the sediment-water interface as a cement. Contrary to Hollman (1962, 1964), who emphasized subsolution, Milliman and Müller thought that precipitation had the dominant role in nodule formation. They did note that aragonite is rare or absent in areas away from pteropod productivity, but they did not discuss the question of aragonite dissolution. This model of nodule formation was recently adopted by Müller and Fabricius (1974) to interpret the paleogeography of the ammonitico rosso; they postulated a restricted marine setting similar to that of the eastern Mediterranean today. This conclusion has been criticized because the ammonitico rosso faunas are open marine. Perhaps a semantic problem is causing misunderstanding. We commonly tend to picture restricted environments as hypersaline or hyposaline lagoons or bays. The eastern Mediterranean in fact supports a good marine fauna; it is restricted only in the sense that the sill at Gibraltar is shallow enough to prevent the inflow of cold Atlantic bottom waters. The restriction is sufficient to make the Mediterranean Sea warmer and more saline than open oceans but not enough to inhibit the growth of normal marine organisms. Why should nodules be formed in a restricted marine environment? Was a geographical restriction similar to the Strait of Gibraltar sufficient to cause the genesis of nodules? Milliman and Müller (1973) provided the answer to the second question: Magnesian calcite precipitation did not take place ubiquitously in the eastern Mediterranean, not even during Quaternary time. Magnesian calcite is common in surficial sediments. The modern conditions are apparently conducive to precipitation, but the mineral is rare or absent in sapropel layers, the last of which was deposited 7,000 to 9,000 yr ago. Obviously, a geographical restriction is a necessary factor but not always a sufficient condition for the genesis of this deep-sea cement. The configuration of the Strait of Gibraltar has not changed significantly during the past 10,000 yr, but the Mediterranean circulations have.

Milliman and Müller related the origin of nodules to the temperature and salinity of the Mediterranean Sea; they wrote

> The magnesian calcite layers in the Ionian Sea show a positive correlation with the occurrence of *Cyclococcolithus fragilis* (Lohmann).... This Coccolithophorid is restricted to a narrow temperature and salinity range (18–27°C, 37.7–38.1‰), ... suggesting that magnesian calcite was not precipitated when one or both of the environmental parameters fell appreciably below the minimum required for *C. fragilis.* Apparently such conditions did not occur during either glacial or interglacial times, but were present during transition periods when colder waters flowed from the western into the eastern Mediterranean with rising sea level, and when glacial meltwater runoff lowered surface salinities.... (1973, p. 48).

A flaw in their reasoning is that the genesis of magnesian calcite should be related to the bottom waters, not to surface or intermediate waters where coccolithophorids lived. Natterer (1899) demonstrated that the lithification was the work of bottom circulation.

The Mediterranean Sea today is separated from the Atlantic Ocean by a shallow sill, and the inflow consists mostly of warm surface waters. The basin would be stagnant if there were no generation of new bottom waters within the Mediterranean. The two major sources of bottom waters at present have been identified in the North Balearic and the Adriatic. The new bottom water from the Balearic travels in a direct path to Gibraltar (Fig. 5). The Adriatic water, chilled by northers during the winter, descends ($T < 11$°C) and flows out of the Adriatic. This bottom-water mass at 1,400-m depth has a temperature of about 13°C and a salinity of 38.6‰ (Lacombe and Tchernia, 1973). In the Ionian Basin, the Adriatic cold water meets the saline Levantine Intermediate Water, and the product of this mixing, water of 14.5°C and 38.83‰, is believed to be the source of bottom circulation of the Tyrrhenian, Ionian, and Levantine Seas (Miller, 1973; see also Fig. 5). The flow out of the Adriatic is probably sporadic, and it diminishes in intensity and distance from origin. The magnesian calcite cement is found in the path of this bottom current in the eastern Mediterranean, and the precipitation is apparently an expression of the interaction between sediments and the moving water mass.

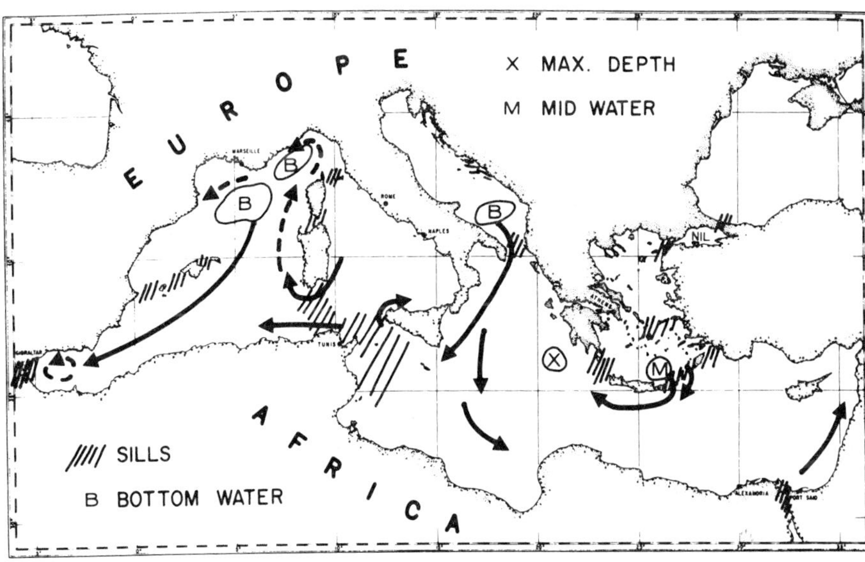

Figure 5. Circulation of Mediterranean bottom currents (after Miller, 1973). Strait of Gibraltar effectively barred penetration of Atlantic bottom waters into Mediterranean Sea. Bottom circulation of eastern Mediterranean was initiated where the saline Levantine Intermediate Water, flowing westward, met supercooled Adriatic Bottom Water. Circulation of this bottom current is believed to be responsible for penecontemporaneous diagenesis of carbonate ooze in eastern Mediterranean basin.

The circulation pattern during glacial times might be similar. The descending water may be less saline because the evaporative losses are smaller, but colder because more severe temperatures prevail. However, the circulation pattern during the transitional period 7,000 to 9,000 yr B.P. was very different. Olausson (1961), Ryan (1969), and Milliman and Müller (1973) all agreed that the sapropel was formed during a stage of bottom stagnation in the eastern Mediterranean. It was pointed out that the continental glaciation during early Holocene time (7,000 to 9,000 yr B.P.) caused large influxes from fresh water: this lowered surface salinities, thereby facilitating density stratification in the eastern Mediterranean. The stagnation resulted when chilled surface water failed to acquire sufficient density to descend and induce bottom circulation in the Ionian Basin. When there was no circulation, there was no cementation, and there was no formation of nodules.

Returning to the question of ammonitico rosso, we could easily perceive the existence of a partially restricted Tethys. The separation of Africa from Europe was not yet final, and the Alpine Tethys was separated from the open ocean to the east by a stratigraphically placed Greco-Italian microcontinent (Fig. 6). One or more shallow seaways permitted the inflow of surface waters, but the sill depth was shallow and therefore barred the intrusion of the bottom currents from the Indian-Pacific Ocean. As in the present Mediterranean, the bottom circulation might have been induced by surface-chilled water within the Alpine-Tethyan basin. The periodic circulation of a bottom current could have been responsible for the origin of nodules and nodular bedding surfaces of the ammonitico rosso facies. I tend to agree with Jenkyns (1974) and with Müller and Fabricius (1974) that the nodules formed from submarine cementation. Nodules grew and coalesced, forming hardground. Dissolution probably did play a role in the genesis of nodular limestone. However, as Schlager pointed out, ammonites in ammonitico rosso are almost invariably internal molds, and this mode of fossilization is indicative of early diagenetic dissolution at a few metres depth beneath the sea floor. We have no clear-cut evidence that submarine dissolution was responsible for nodular growth. It should be recalled that stratigraphic hiatus and supposed surfaces of submarine dissolution are disconformable contacts between beds of nodular limestone; the nodular limestone between such disconformities may represent periods of growth and deposition, rather than dissolution. In fact, I am not even certain if nondepositional intervals resulted from subsolution by chemical processes or from nondeposition (above a hardground) because of bottom-current circulation. Returning finally to the question of red color, I suspect that the color does not have much to do with a distant spreading ridge. Rather, the bottom current was a water mass freshly descended from the surface, and it was well enough oxygenated to cause extensive oxidation of iron minerals in areas of relatively slow sedimentation.

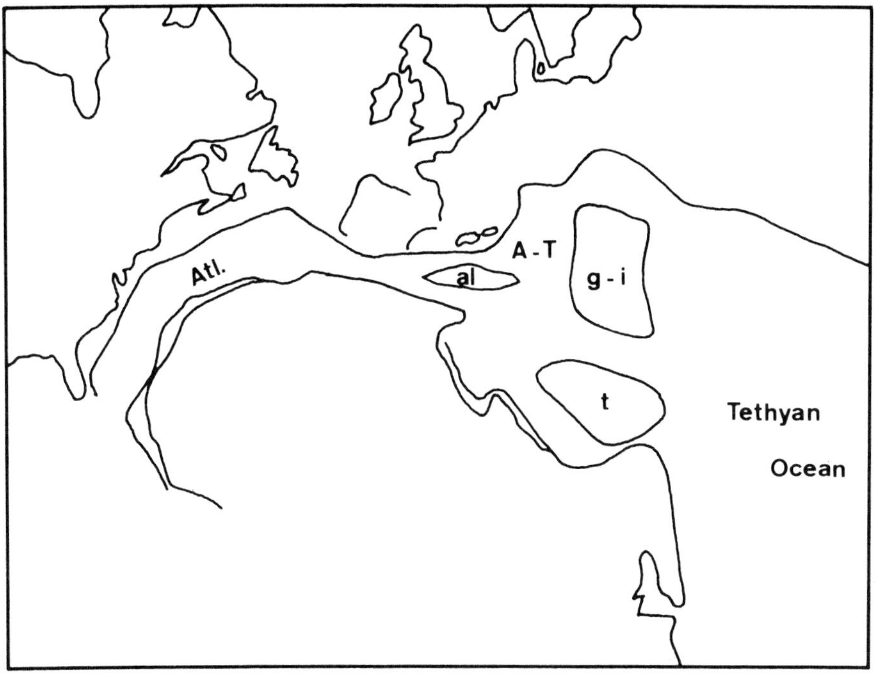

Figure 6. Paleogeography of Middle Jurassic Alpine Tethys. Relative positions of Africa, Europe, and North America during Middle Jurassic time have been determined by Dewey and others (1973) on basis of magnetic lineations in Atlantic Ocean. Possible position of Turkish (t), Greco-Italian (g-i), and Alboran (al) microcontinents is speculative. Map shows that Atlantic (Atl) was then a narrow ocean, and that Alpine Tethys (A–T) was an island sea, probably separated from Tethyan Ocean by shallow sills, which prevented intrusion of oceanic bottom currents.

Depth of Late Jurassic Tethys

The depth of the Late Jurassic Tethys sea has been a subject of controversy for 100 years. Here I will explore the problem from the point of view of the apparent sequential reversal of the Alpine pelagic sediments.

The phenomenon of dissolution cycles was first noted during the JOIDES III drilling in the South Atlantic. The basal formation directly above the basalt is always a calcareous ooze that has been subjected to slight dissolution (eolytic facies). This formation is succeeded by calcareous oozes showing increasing signs of more and more advanced dissolution (oligolytic and mesolytic facies). The final sediment of the cycle is red clay in which all carbonate material has been removed by dissolution (hololytic). This sequence has been considered an indication of ocean-floor subsidence as the ocean crust moved laterally from an active spreading ridge to the deeper ridge flank (Hsü and Andrews, 1970). In the Pacific basin, where the accumulation of siliceous oozes has been significant, an intermediate siliceous facies may intervene between the carbonate sediments and the red clay in a dissolution cycle (Fig. 2). The concept of silica compensation depth was adopted to explain the absence of siliceous fossils in pelagic clay (Bukry and others, 1970). The deep-sea facies pattern is complicated if a site moved in and out of the equatorial belt of high production during its spreading history. Without such complication, the normal stratigraphically ascending sequence for a gradually deepening crust would be basalt-limestone-radiolarite-red clay (see, for example, Heezen and others, 1973). This sequence was indeed recorded at many Pacific drilling sites; veterans on the *Glomar Challenger* could almost sense the approach of the basalt basement as the drill string punched through chert and siliceous ooze into a calcareous sequence. Yet the Alpine Tethys sequence is apparently reversed: where the sedimentary contact is visible, the first sediment overlying the ophiolite is almost invariably radiolarite of Late Jurassic age, overlain by pelagic limestone mainly of Early Cretaceous age. A more or less coeval radiolarite-limestone sequence is found in many parts of the southern Tethyan margins (Bernoulli, 1972; see also Fig. 3). What is the significance of this apparently anomalous sequence?

Fuchs, Steinmann, Fischer and Garrison, and others all postulated radiolarite deposition beneath the calcite compensation depth. Because this depth ranges from 4,500 to 5,500 m, a figure of 5,000 m has been postulated as the possible depth of the Jurassic Tethys (for example, Model 2 of Fischer and

Garrison, 1967). Fischer and Garrison (Model 1) did consider also the possibility that the calcite compensation depth was shallower in Late Jurassic time than in Early Cretaceous time, thus accounting for the radiolarite-limestone succession.

Were the Alpine radiolarites deposited below the calcite compensation depth? Did the Jurassic Tethys have a calcite compensation depth similar to that of the modern open ocean? In radiolarite completely devoid of $CaCO_3$, one may agree with Steinmann that those sediments were deposited below the calcite compensation depth. However, some radiolarite beds have calcite aptychi, and the Radiolarite Formation is commonly associated with the Aptychus Limestone. Studying the detailed dissolution phenomena, Schlager (1974) concluded that the two formations, although clearly representing different facies, display the same preservation of cephalopods, apart from the generally greater abundance of these fossils in the aptychus beds. Because the calcite aptychi are for more abundant than the aragonitic phragmocones, one might conclude that aptychus-bearing sediments were deposited above the calcite compensation depth, but below the aragonite lysocline. Calcite-free radiolarites were sometimes deposited beneath the calcite compensation depth when it was unusually high. So we could postulate rapidly fluctuating calcite compensation depths but not bouncing sea floors.

The concept of lysocline was introduced by Berger (1967) to denote a zone of rapidly increasing solution rate above the compensation depth. Murray and Renard (1891, p. 226–228) noted that pteropod ooze is rare below 1,400 fm (2,600 m), and we might adopt this figure as the approximate level of aragonite lysocline in the open ocean today, although the aragonite compensation depth in the Atlantic has been placed at some 4,000 m (Chen, 1964).

In an inland sea like the eastern Mediterranean, aragonite dissolution does not seem to correlate with depth. Milliman and Müller (1973) found almost no aragonite in all but 3 of their 21 cores. Those cores were taken from depths ranging from 1,000 to 4,000 m. The absence of aragonite was attributed to nonproductivity, although subsolution may have played a role, especially in regions of hardgrounds. Paradoxically, the three cores that did contain appreciable amounts of aragonite were located in the central abyssal-plain province, one at about 3,000 m and the other two at 4,000 m! The aragonite occurs within the lutite fraction, and its preservation was apparently not related to high pteropod productivity. Rather, those central basin areas seemed to have stayed out of the path of the aragonite-dissolving bottom currents. A depth estimate of lysoclines in ancient environments is further complicated by the possibility of diagenetic dissolution. As mentioned previously, pteropod shells were dissolved after burial in some Red Sea sediments in water less than 1,000 m deep.

I have concluded that the Middle Jurassic Tethys was similar to the present eastern Mediterranean and that the ammonitico rosso may have been subjected to extensive diagenetic dissolution of aragonite. It would thus be very foolhardy to assign any depth figures for the ammonitico rosso sea. However, a significant change in paleogeography was to follow. The Mediterranean, with its nutrient-poor waters, does not support a healthy radiolarian fauna. In contrast, radiolarians flourished in the Late Jurassic Tethys, which was apparently part of the open oceans. We also do not need to worry too much about diagenetic dissolution of aragonite, because phragmocone casts, which are so very abundant in the ammonitico rosso, are nearly absent in the radiolarite and Aptychus Limestone. Thus, it would not be entirely misleading to estimate the Tithonian water depth on the basis of lysocline and calcite compensation depth.

It should be emphasized that both lysoclines and compensation depths lie well below the level of saturation; sea water is undersaturated not only for aragonite but also for calcite at depths of a few hundred metres. Peterson (1966) concluded, on the basis of experimentation with suspended calcite spheres, that carbonate sediment is being dissolved as soon as it falls below the zone of saturation. The dissolution rate is, however, very slow at shallower depths but increases sharply at depths of 3,000 and 3,700 m. The zone of rapidly increasing dissolution rate has been defined as the lysocline (Berger, 1970b). Tiniest and most susceptible particles will dissolve totally on their way down; only the most resistant or faster sinking large particles have a chance to fall through the lysocline and sink beneath the compensation depth. If the calcite production was relatively minor and if the product consisted of tiny and nonresistant nannofossils, we might well suppose that some of those slowly descending particles are completely dissolved even before they reach the bottom. Those that reach the sea floor could be completely removed where bottom circulation is effective in subsolution.

Whether the ammonite phragmocones were dissolved on their way down or after they reached the bottom cannot be determined. In either case, their removal strongly suggests that the sea-floor depth should lie within or below the zone of rapid aragonite dissolution. The preservation of aptychi suggests a depth somewhat above the calcite lysocline. At present, such a depth should range from 2,600 to 3,300 m. The present calcite compensation depth is on the whole several hundred metres deeper than the Tertiary and Cretaceous depths (Berger and von Rad, 1972), and the present lysoclines are perhaps deeper by the same order of magnitude. If we add a correction factor of −500 m for the Late Jurassic lysoclines, we might conclude that large parts of the Tethys were then 2,100 to 2,800 m deep, averaging some 2,500 m when aptychus-bearing radiolarite was deposited.

Some radiolarites are free of calcium carbonate, but they are closely associated with aptychus-bearing beds. Instead of postulating a rapidly changing bathymetry, one could imagine a spatial or temporal variation in calcite compensation depth. We might postulate that the calcite compensation depth in the Late Jurassic Tethys was locally 2,500 m or less where or when carbonate production was modest, or where strong bottom currents effected extensive subsolution. We need not go to the extreme of assuming repeated bathymetric changes of the order of a thousand metres or more to account for the radiolarite-aptychus association.

A completely novel approach to determining paleobathymetry has been made possible through geophysical considerations of mantle density and isostatic equilibrium. Empirical observations have shown that all active spreading ridges have a similar crestal elevation and that there is a uniform relation between the depth and age of the oceanic crust. The newly created crust under the East Pacific Rise in the North Pacific has an elevation of 2,800 m. Mid-ocean ridges elsewhere, with the exception of the Mid-Atlantic Ridge south of Iceland, have generally shallower crestal elevation, but the differences are less than 400 m in most cases (Sclater and others, 1971; Sclater and Detrick, 1973). Regions of older crusts are underlain by mantle of higher density and thus have a lower elevation or greater depth. Within 50 m.y. the basalt sea floor would be at a depth of 5,000 m or more. The final stable equilibrium would be reached after 150 m.y., when a bare basement would be at a depth of just a bit less than 6,000 m. The depth figures are to be adjusted slightly if there are sufficiently thick sediments to cause appreciable isostatic subsidence. Sclater and Francheteau (1970) presented a theoretical model, utilizing heat-flow data, to account for the subsidence. Later, Sclater and Detrick (1973) convincingly proved the validity of this theory through an analysis of JOIDES drilling data (Fig. 7). Their reasoning is so logical and the empirical evidence is so overwhelming that there should be no reason to doubt that the newly created sea floor should have a certain limited range of elevation. The crestal height has remained (since Cretaceous time at least) between 2,400 and 2,800 m (Sclater and Detrick, 1973). The Pacific, the largest of all oceans, has the deepest crestal elevation; new crust in a small ocean should have a depth on the shallower side of the range. A round number of 2,500 $\pm$ 200 m would represent the most reasonable estimate for the elevation of a newly created Tethys floor.

Most of the Alpine Tithonian radiolarite of the ophiolite zone lies directly above a newly created sea floor. Geophysical considerations would thus give 2,500 $\pm$ 200 m as the depth for the central Tethys in early Tithonian time. In this relatively narrow intercontinental ocean, the newly created crust probably did not occupy a ridge position. We might conclude that the radiolarites on the southern Tethys margins were deposited in a somewhat shallower depth, as is true in the Red Sea, where the newest crust underlies an axial depression.

Although two independent lines of evidence—paleoceanographic and geophysical—should both yield the same estimate of about 2,500 m, more reliance should probably be put on the latter. We should keep in mind the fact

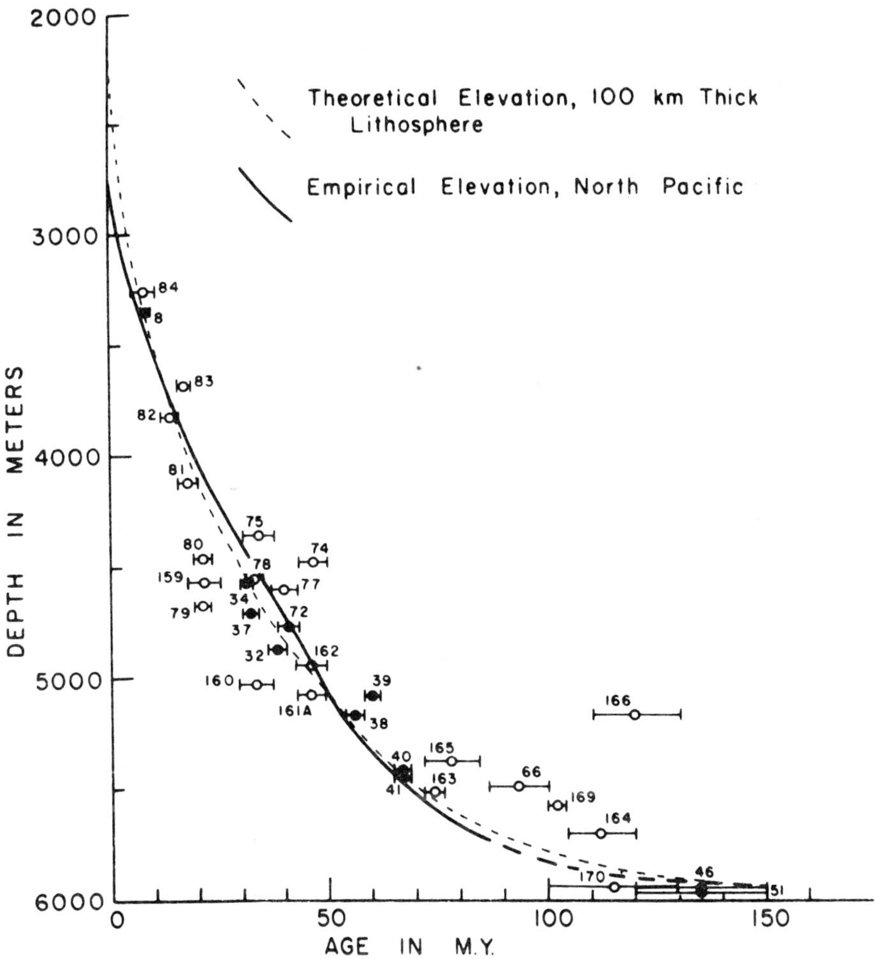

Figure 7. Subsidence of aging crust (after Sclater and Detrick, 1973). Sclater and his co-workers have found, on the basis of geophysical considerations, that newly created oceanic crust has an elevation of 2,600 ± 200 m. With cooling of crust as it is spread away from an active ridge, ocean floor sinks. This figure shows that theoretical prediction has been confirmed by results of JOIDES deep-sea drilling. Theoretical elevation from Sclater and Francheteau (1970); empirical elevation from Sclater and others (1971).

that Aptychus Limestone and calcareous radiolarite are associated with carbonate-free radiolarite and red clay (see Cadisch and others, 1968) that were deposited beneath the calcite compensation depth. In Late Jurassic time there must have been considerable variation in calcite compensation depth, perhaps from a depth similar to that of the present oceans to one of less than 2,500 m.

If we accept the geophysical arguments, we must accept the implication that the Alpine Tethys was not deepest during the time of Late Jurassic radiolarite deposition; large parts should have been deeper millions of years after the initial spreading and emplacement of ophiolites. Colom (1958), working in the Subbetic zone of Spain, noted the existence of globigerinids in Tithonian sediments and their absence in Lower Cretaceous sediments, and he came to the conclusion that the Alpine Tethys did not reach its greatest depth until Neocomian time, sometime between the Valanginian and Hauterivian Stages, when some 20 or 30 m.y. had elapsed since the initial appearance of ophiolite. The extensive dissolution of planktonic Foraminifera suggests that the Early Cretaceous Tethyan sea floor might be placed below the calcite lysocline, at depths of 3,000 or 4,000 m. The deeper estimate of 4,000 m would be more in line with the prediction based on the "Sclater curve" (Sclater and Detrick, 1973; see also Fig. 6). The Cretaceous sequence (Maiolica-Palombini-Lavagna) in the ophiolite zone of the Ligurian Apennines provides another indication of gradual crustal subsidence during Early Cretaceous time. The Maiolica is a pelagic deposit, full of nannofossils. The middle Cretaceous Lavagna is a black shale devoid of carbonate. The Lavagna facies has a parallel in the coeval dark clays of the central Atlantic; they were all deposited after the sea floor again sank beneath the calcite compensation depth (Bernoulli and Jenkyns, 1974, p. 146), which should have been 4,000 to 5,000 m deep, comparable to the present calcite compensation depth. If this analysis is correct, the Alpine radiolarite-limestone represents a sequence of increasing, not decreasing, depth. We must find another reason for the facies change from siliceous to calcareous sedimentation.

Trümpy (1971) noted the great horizontal spread of the radiolarian chert-Aptychus Limestone suite in the Tethys and estimated a minimal north-south extension of 250 km and an east-west extension of 1,200 km. He pointed out the inescapable implication that an oceanlike body of water came into existence during Late Jurassic time. An even broader Tethys was indicated by a plate tectonics reconstruction that shows a latitudinal separation of some 10° (1,000 km) between Europe and Africa (Francheteau, 1970; see also Fig. 8). Microcontinents were scattered and largely submerged in this broad seaway and could no longer hinder the free circulation between the Alpine Tethys and the open ocean; the Tethys became a part of the ocean. Plate tectonics analysis and paleomagnetic data further permitted Francheteau to draw paleolatitudes, which showed that the eastern entrance to the Late Jurassic Tethys lay almost at the equator. Equatorial currents from the Pacific would be deflected by Africa to flow along the southern Tethys margins in a direction slightly north of west. It is a well-established principle that areas of high organic production in the oceans are found in the subarctic, subantarctic, and equatorial regions and in areas of coastal upwelling. The present distribution of siliceous oozes is largely coincident with those areas of high production (Riedel, 1963; Calvert, 1974). I am tempted to postulate that the Late Jurassic radiolarite represents the biogenic deposit from an equatorial current. Bernoulli (1972) noted a simultaneous first appearance of radiolarite in western Greece, the central Apennines, and the southern Alps during Callovian time. The onset of the radiolarite sedimentation coincided obviously with a tectonic event that removed the obstruction barring the entrance of the equatorial current, perhaps the same obstruction that restricted the Tethyan sufficiently to induce the deposition of the nodular limestones. According to this line of reasoning, the facies change from ammonitico rosso to radiolarite was not caused by a sudden

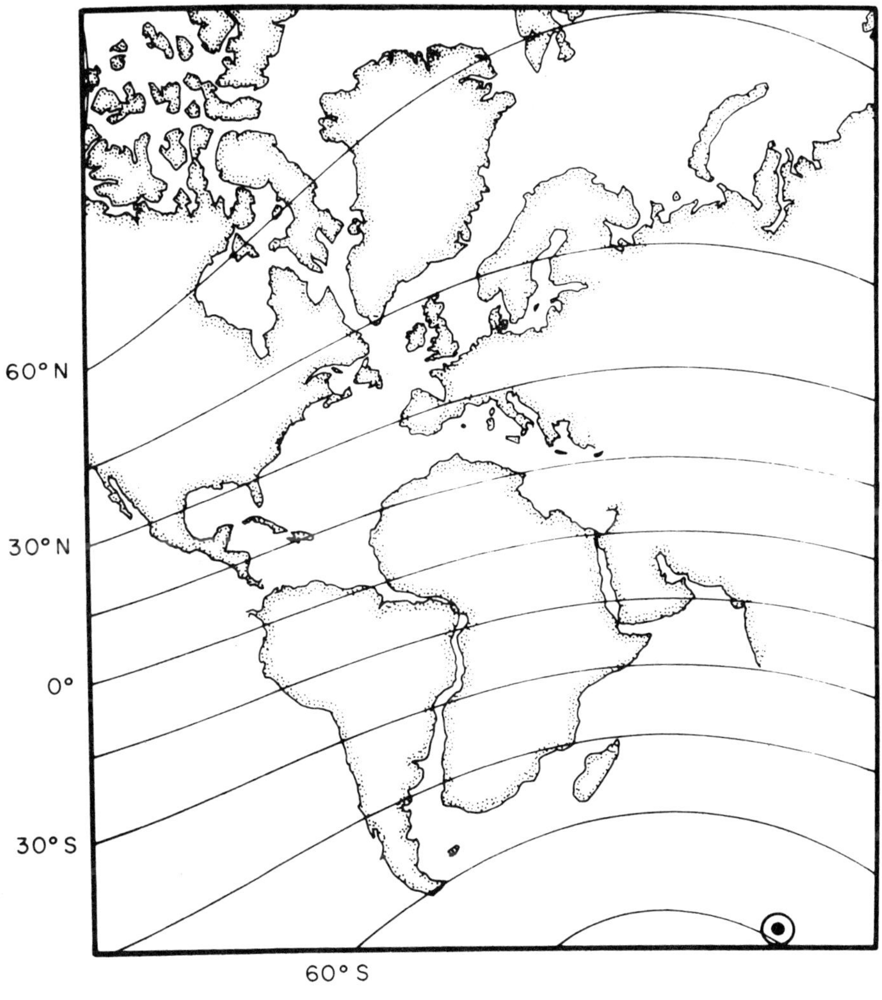

Figure 8. Paleogeography of Late Jurassic Tethys, 130 m.y. ago (after Francheteau, 1970). Alpine Tethys became broad seaway during Late Jurassic time and lay just north of Jurassic equator. Equatorial currents from east might be deflected northward by Africa and bring nutrients for radiolarian blooms in Tethyan geosyncline.

and synchronous subsidence that depressed the Tethyan sea floor beneath the calcite compensation depth. Instead, the facies change responded to a critical paleographical change that reorganized the circulation patterns of the world's oceans. The stratigraphic record suggests that the equatorial current did find its way to the Austrian Alps during Middle Jurassic time (Garrison and Fischer, 1969), to eastern Greece during Early Jurassic time (Aubouin and others, 1963), and to Turkey in Late Triassic time (Dumont and others, 1974). Apparently, the Pacific equatorial current was in existence during all these times, but it did not invade the western Tethys until Callovian time. Not all Late Jurassic Alpine Tethyan pelagic formations are radiolarite. Red nodular limestones of the Oxfordian, Kimmeridgian, or Tithonian Stages are present locally in northern Italy (rosso ad aptici), in Sicily, and in the Balkans (Sapunov and Ziegler, 1975). The Late Jurassic sediment encountered in the western Atlantic JOIDES Leg XI drill sites also belongs to the ammonitico rosso facies (Bernoulli, 1972). Apparently, the radiolarian sediments were restricted to the central Tethyan belt where open-ocean conditions prevailed. At the same time, nodular carbonate sediments were formed in peripheral basins, outside the path of this Late Jurassic equatorial current system.

The change from Middle Tithonian radiolarite to Late Tithonian limestone sedimentation poses a more difficult problem to resolve. We could hardly postulate a major shift in equatorial current, as the production of calcareous plankton also requires nutrients. Perhaps we should consider the possibility that the silica consumption was so great during deposition of radiolarite that the world's oceans became seriously depleted in silica.

Maiolica and Chertification

A simple calculation might dismiss the idea that the Early Cretaceous ocean was depleted in silica because of the Late Jurassic population explosion of Radiolaria in the Alpine Tethys. If we take Trümpy's (1960) conservative estimate of the dimension of the Tethys in Tithonian time and assume an average thickness of 40 m for the radiolarite bed, the total volume of silica produced would be $1.2 \times 10^8 \times 2.5 \times 10^7 \times 4 \times 10^3 = 1.2 \times 10^{19}$ cm^3, or 3×10^{19} g (assuming a density of 2.5 g/cm^3). If this amount of silica was deposited in 10 m.y. of the late Jurassic Period, the annual consumption should have been 3×10^{12} g/yr. If the Jurassic supply of silica to the open ocean from weathering is as great as at present and amounted to 4.3×10^{14} g/yr (Livingston, 1963), the Tethyan consumption would represent only 1 percent of the supply. Even if we assume a somewhat exaggerated immense Alpine Tethys of 3,000 km by 1,000 km, the estimated Tethyan consumption would still be less than 10 percent of the supply. It seems absurd to assume that the silica sedimentation in such a local basin could appreciably deplete the silica content of the ocean.

However, Tithonian radiolarian chert is known from other mountain belts (see Grunau, 1965). More important is the fact that the Pacific equatorial belt of high silica production existed in Late Jurassic and earlier times, as shown by the JOIDES drilling (Fischer and others, 1971; Heezen and others, 1973). This equatorial belt is and has been a major consumer of silica in the ocean. We might assume that there has been a long-standing steady state, when the ocean, as a great buffer, deposited as much silica as it received. However, the sudden increased demand by the Alpine Tethys in Late Jurassic time could temporarily upset the delicate balance and induce a silica crisis.

Berger (1970a) and Calvert (1974) made an estimate of the oceans' silica budget today: (1) total amount of silica in oceans today, 11.1×10^{18} g (Goldberg and Arrhenius, 1958); (2) influx from weathering, 4.3×10^{14} g/yr (Livingston, 1963); (3) biogenic extraction from oceans, >1.7 to 3.2×10^{16} g/yr (Heath, 1974), >8 to 16×10^{16} g/yr (Lisitsin, 1971); (4) return to ocean from bottom sediments, $0.7 \times 7.9 \times 10^{14}$ g/yr (summary by Calvert, 1974); (5) rate of deposition assumed to equal influx from weathering, 4.3×10^{14} g/yr. Some of the estimates vary by about an order of magnitude and are therefore quite uncertain. However, several conclusions have been drawn by these authors: (1) Silica influx into the ocean could supply only about 1 percent of the silica needed for biogenic production. (2) Silica returning from bottom sediments into the ocean across the sediment-water interface is just as important a source of silica as the influx from weathering, which apparently is the only important source on land. (3) If a steady state is assumed, only a few percent of the biogenic silica extracted from the ocean ever reached the sea floor to be preserved; much of the biogenic production must have been redissolved during the descent through the water column. What those authors did not mention was a startling implication that if the depositional rate should exceed the influx from weathering by 1 percent, say 4.3×10^{12} g/yr, the silica in the ocean water should be completely depleted in 3 m.y. Of course, I am speculating like a student estimating the population growth. The ocean acts as a buffer: if the amount of silica in the ocean is becoming depleted and if there is no accelerated supply from land, there should be less production, increased dissolution, and less deposition. In terms of stratigraphy, less deposition means a retardation of radiolarite sedimentation. When this process is carried to an extreme, silica sedimentation might cease locally altogether, and a radiolarite facies would be replaced and overlain by another pelagic facies, such as a pelagic limestone.

Returning to the question of Radiolarite-Maiolica succession, one might reasonably assume that the silica budget was in a steady state during Middle Jurassic time; all the silica brought into the ocean would be deposited as oceanic deposits, mainly in the equatorial zone. At the beginning of Callovian time, the nutrient-rich equatorial current first swept into the Alpine Tethys, inducing the widespread silica sedimentation there. We have shown that if the deposition in the Tethys represented the net loss of silica from the ocean, its silica reserve would have been depleted in 4 m.y. The ocean, as a buffer, cannot permit this extreme state; silica consumption must be conserved. What would be more logical than to assume that the Alpine Tethys, located at the end of a westward-flowing equatorial current, would come up short after the start of silica shortage? Radiolarian bloom would come at less frequent intervals as silica became depleted in the current, but other nutrients would still be there. The fertile "soil" would be taken over by a new race of plants, the calcareous nannoplankton. Thus it seems to me not too farfetched to interpret the Radiolarite-Maiolica contact as the record of a silica crisis in the ocean and to hypothesize that the decline of Radiolarite permitted the explosive evolution of the coccoliths during latest Jurassic and Early Cretaceous time. The silica crisis should not have been of much concern to the major consumers; the equatorial bulge continued to get its share, as shown by the JOIDES drilling results. However, the pinch was probably felt by fringe consumers. Having worked for a decade in the California Coast Ranges, I have been puzzling over a fact, recently confirmed by Pessagno (1973) that few Franciscan radiolarian cherts could be definitely assigned an age younger than Tithonian. Could this be the result of the same crisis? A consequence of increased production of calcareous nannofossils is that the calcite compensation depth is lowered. We have shown that this depth was at times less than 2,500 m during Late Jurassic time and probably was more than 4,000 m at the end of Early Cretaceous time in the Alpine Tethys. This lowering of the calcite compensation depth permitted the preservation of calcareous sediments on a subsiding sea floor, together with radiolarian debris, leading to the accumulation of the Maiolica siliceous limestone. The paradoxical radiolarite-limestone transition in the Alpine Tethys is an apparently regional occurrence related to regional production and dissolution of planktonic organisms; we cannot make a generalization that geosynclinal sequence is always the reverse of that of a spreading ridge.

A second remarkable aspect of the Maiolica facies is the common occurrence of chert nodules or lenses. There is considerable evidence that the silicification was early diagenetic, prior to the lithification of the host rock, and this interpretation is confirmed by the finding of chert nodules in unconsolidated nannoplankton oozes of the Maiolica facies in JOIDES hole 100 (Bernoulli, 1972).

A great deal has been learned about chert genesis since the recovery of chert by deep-sea drilling (Garrison, 1971; Heath and Moberly, 1971; Berger and von Rad, 1972; Heath, 1973; Lancelot, 1973; Wise and Weaver, 1974). The following observations have been made:

1. Chertification is not an automatic process of biogenic siliceous sediments. In a 100-m-thick sequence of siliceous sediments, one might encounter only a few metres of layered or nodular chert; the rest is unconsolidated ooze. The contact between a hard chert and a soft ooze is clearly defined (Maxwell and others, 1970).

2. Chert may replace pre-existing calcareous sediment, and silicification of lime turbidite is a common occurrence (Garrison, 1971).

3. The youngest chert reported has a late Miocene age, in a region of steep geothermal gradient (Heath, 1973).

4. Two forms of silica have been identified: cristobalite and quartz (Heath and Moberly, 1971; Berger and von Rad, 1972; Lancelot, 1973). Lancelot (1973) believed that "chert occurring in clayey layers (zeolite clays, clayey radiolarian oozes, and marly limestones) is predominantly disordered cristobalite, and . . . chert in carbonate sediments is predominantly quartz." Wise and Weaver (1974), however, found exceptions to this rule.

5. Cristobalite has been found to associate with clinoptilolite (Rex, 1970; Gibson and Towe, 1971; Wise and Hsü, 1971), but the association is not universal (Wise and Weaver, 1974).

6. Occurrence of chert is not always predictable: sediments apparently devoid of any siliceous organisms may contain abundant chert (Lancelot, 1973); the association of chert with volcanogenic sediment has been postulated (Gibson and Towe, 1971), but some of such sediments are conspicuously devoid of chert (Lancelot, 1973).

7. Calcareous turbidite associated with siliceous ooze is commonly chertified (Beall and Fischer, 1969), and this generalization seems to be confirmed by observations on land (Nisbet and Price, 1974).

Considerable empirical observation has been made, but the environments that are conducive to chertification are still not known, because no chert has been reported from the Holocene marine environment. Concerning the apparent absence of Holocene chert, Lancelot (1973, p. 400) commented "(1) Chert is not presently forming because environments existing are now different from the one existing during the early Tertiary time; [or] (2) Formation of chert is such a slow process that it has not had enough time to form . . . ." Although Lancelot favored the first alternative, he was at a loss to describe what could be the difference between the Holocene environment and that of the past. One might add that chertification requires an efficient mechanism of silica transfer through sediment. Is it possible that the Holocene ocean-bottom environments are not conducive to silica transfer? Two possible mechanisms of material transfer were responsible for diagenesis: ground-water movement or diffusion through the pore fluid. In the essentially hydrostatic deep-sea environment, ground-water transfer of silica can be ruled out; we have to rely on diffusion across a concentration gradient. Studies of pore waters in Holocene sediments have shown that concentration gradients indeed exist near the sediment-water interface. Bischoff and Sayles (1972) found three types of silica concentration gradient (Fig. 9). The first is characterized by an initial increase from about 10 ppm at the sediment-water interface to about 20 or 30 ppm at 1- or 2-m depths and greater; this is the general pattern found in the Atlantic samples (Bischoff and Ku, 1970) but in only a few of the cores from the East Pacific Rise (Bischoff and Sayles, 1972). The second is characterized by a continuing and somewhat erratic increase to concentrations of 40 or 60 ppm at 7- or 9-m depth. The third and most common pattern on the rise is characterized by a rapid increase from 10 to 20 ppm at the sediment-water interface to a maximum of some 50 ppm within the first metre or two, followed by a gradual decrease to a lower steady value of 20 to 30 ppm. The existence of the silica concentration gradients can be traced to the differential solubility of amorphous silica and quartz. The solubility of amorphous silica is 140 ppm at 25°C in pure water and in sea water (Krauskopf, 1956; Siever, 1957), but the solubility is considerably less at the lower temperatures prevailing on the ocean bottom; the solubility of amorphous silica at 1° to 3°C is probably 60 to 80 ppm, as suggested by the maximum values for SiO_2 in interstitial water squeezed out of siliceous sediments under in situ conditions (Fanning and Pilson, 1971; Bischoff and Sayles, 1972). In contrast, the silica content of the ocean bottom water is similar to that of quartz solubility; it has been estimated to range between 6 and 14 ppm at 25°C (Siever, 1957) and is probably 10 ppm or less under ocean-bottom conditions. Any type of gradient would transfer silica from the bottom sediment to the overlying bottom water through the near-surface zone of mobility, where a concentration gradient exists. The silica diffusion is, however, insignificant within the deeper zone, where the pore water has a relatively constant concentration. The lower boundary of silica mobility would be 1 or 2 m for the first and third types and 7 or 9 m for the second type. Assuming a steady-state concentration profile and an average sedimentation rate of 2 cm/10^3 yr, a silica-bearing sediment would be out of the zone of silica mobility after 50,000 or 100,000 yr in the first and third types and after somewhat less than 500,000 yr in the second type.

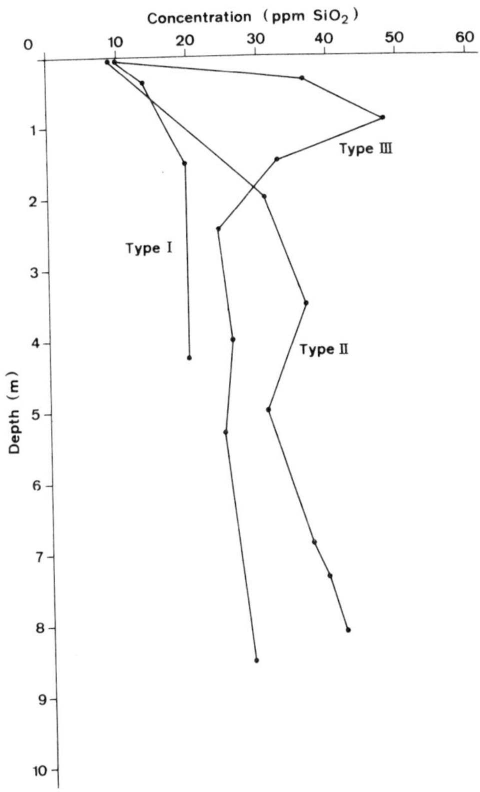

Figure 9. Concentration of silica in interstitial waters (data from Bischoff and Sayles, 1972).

The flux of silica across a sedimentary column vertically upward is the product of coefficient of diffusion and concentration gradient. Wollast and Garrels (1971) determined the diffusion coefficient for silica in sea water to be 1.0×10^{-5} m^2/sec at 25°C. Bischoff and Sayles (1972) estimated that the value should be decreased to about 1.5×10^{-6} cm/sec for present ocean-bottom conditions in 30 percent diffusional porosity sediments. We can thus calculate the flux and the total silica able to move through a sediment column before it is depressed beneath the zone of mobility. The most favorable type (involving most silica transfer) is the third. With an optimum gradient of 40 ppm/m or 4×10^{-7} g/cm³/sec, the diffusion flux is 6×10^{-13} g/cm²/sec, or about 2×10^{-5} g/cm² yr. Assuming 50,000 yr should elapse before the top of the present zone of mobility is buried, the total amount of silica transfer would be 1 g/cm² across the 1-m-thick interval. If all the silica, transferring across the zone of mobility, were somehow precipitated as quartz on the sea floor, the resulting chert layer could be 2 to 3 cm thick. In regions of very slow sedimentation, say 1 mm/yr, the resulting chert layer could be twenty times thicker. The chertification process should have been even more efficient during the past, when ocean bottom waters were warmer and the diffusivity coefficient considerably greater. In any case, the existence of a silica gradient beneath the sea floor permits a theoretical possibility of penecontemporaneous chertification. However, coring and dredging during the past century failed to reveal any significant chertification of the contemporary ocean floor. The silica transferred across the zone of mobility is generally believed to be released to the ocean water to make up for the silica depletion caused by biogenic overproduction.

There is no apparent reason for not finding Holocene chert in the deep sea. We might at this stage search for some answers through a review of chert occurrences in ancient rocks. Our computations have shown that chert should most commonly be found in regions of very slow sedimentation if chertification were related to silica transfer across the zone of silica mobility directly beneath the sea floor. As a matter of fact, chert is commonly found in ancient calcareous turbidite in a siliceous ooze sequence, and that turbidite must have been deposited at a very high rate. The shape of chert masses gives us a second clue. It is well known that chert does not form laminar encrustations that are morphologically typical of chemical precipitation in regions of slow sedimentation. Commonly, chert occurs as lenses or nodules. Such nodules are usually ellipsoidal, and along their outermost surface the silica concentration was the same, being equal to the solubility of chert (be it quartz or cristobalite). In other words, the equipotential surfaces of silica concentration potential were ellipsoidal and not horizontal; concentration must have varied not only with depth, but also in horizontal directions. The equipotential surfaces of the zone of silica mobility beneath the present sea floor are horizontal. Silica transfer across the zone could not be responsible for the genesis of nodules. Thus, we must postulate chertification in biogenic sediments after their burial. We could, then, visualize silica transfer from all parts of the enclosing medium across concentration gradients normal to the surface of the ellipsoid; quartz or cristobalite was deposited on the outer surface of the chert where the silica concentration of the fluid was in equilibrium with chert. Nodules grew, and some eventually coalesced with neighbors to form lenticular layers. The aspect ratio (major versus minor axis) of an ellipsoid may give a measure of the relative gradients in the vertical and horizontal directions (Fig. 10).

The formation of chert depends upon the existence of a steady-state

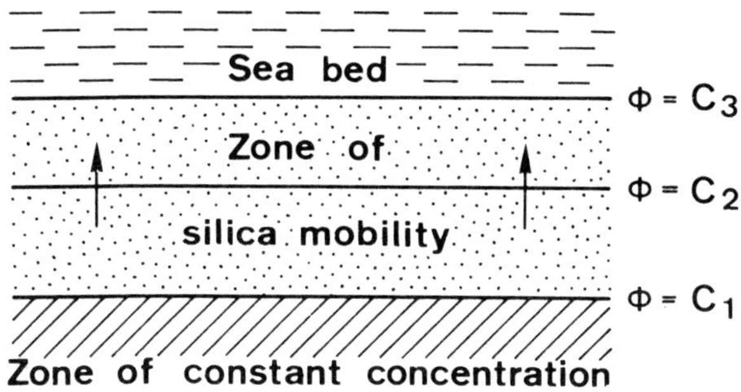

Case I: $\frac{\delta \Phi}{\delta \zeta} = Z = \text{constant}$

$$\frac{\delta \Phi}{\delta X} = X = 0$$

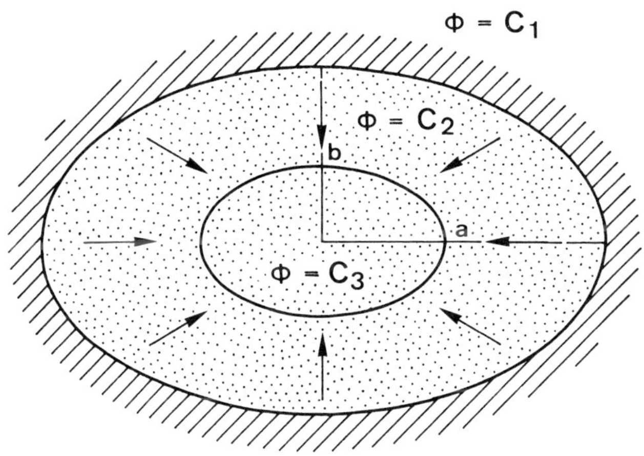

Case II: $\frac{X}{\zeta} = \frac{a}{b}$

Figure 10. Two models of chert diagenesis. Case I. Surfaces of equal concentration potential (ϕ) are horizontal. Silica moves through zone of mobility and theoretically could be deposited as laminations on sea floor. Dredging and coring have failed to obtain sample of penecontemporaneously formed chert, and it is generally assumed that silica moving out of sediments is not deposited, but is released to sea water. Case II. Surfaces of equal concentration potential are ellipsoidal around nucleus. Concentration gradient should be set up between matrix, where biogenic amorphous silica is being dissolved, and outer surface of growing chert nodule, where quartz (or cristobalite) is being precipitated. This mode of chertification may explain genesis of chert nodules in ancient rocks.

gradient for a sufficiently long time to permit the transfer of silica. The difference in solubility between amorphous silica and quartz is probably the cause of the existence of such a gradient. Supply of amorphous silica from biogenic debris could keep the silica in pore fluid at a maximum, but the maintenance of a potential minimum depends upon the continued extraction of silica to deposit quartz (or cristobalite). We have seen that chertification is not an aging process. The Eocene radiolarian oozes I studied during the JOIDES III cruise show no evidence of recrystallization; amorphous silica could apparently be preserved in a metastable state indefinitely. The key to the question is probably nucleation and crystallization kinetics. The prior existence of numerous centers of nucleation is probably necessary to serve as a potential minimum to attract silica by diffusion, and continued precipitation around the nuclei is necessary in order to maintain the concentration gradient indefinitely for chertification. Theoretically, once the process is started, there should be nothing to stop it until all the amorphous silica is converted to chert. The diffusion rate depends upon the concentration gradient, which is in turn a function of the distance between amorphous silica (source) and chert (sink). Effective chertification depends upon a good mixing of quartz debris and siliceous skeletons. The common occurrence of chert nodules in calcareous turbidite might well be related to the mixing of biogenic silica and terrigenous quartz in such sediments. A similar situation could be encountered in other host rocks, and we could therefore not find one particular environment as the one and only environment for chertification.

According to this model of diagenetic chertification, chert, as a rule, has been formed after burial. Burial increases environmental temperatures, facilitates diffusion, and accelerates growth. It is perhaps not a coincidence that the youngest chert (late Miocene) has been found so far in regions of high heat flow and steep geothermal gradient. Apparently even in those favorable environments, chertification must have proceeded for millions of years before nodules could attain respectable sizes.

Concerning chert mineralogy, Lancelot's (1973) postulate of an apparent host-rock control cannot be lightly dismissed. I think the key lies in the mineralogy, rather than in the permeability of the host rock. Bischoff and Sayles (1972) discussed the possibility of sepiolite ($Mg_2Si_3O_8 \times (H_2O)_n$) as a buffer for silica in a pore water: dissolution of skeletal amorphous silica would render a pore fluid supersaturated with sepiolite, resulting in the precipitation of sepiolite around and armoring of skeletal grains. The resulting pore fluid would have a concentration of 30 to 40 ppm. It is possible that cristobalite, not quartz, was precipitated from a pore fluid with such high silica concentration. The common occurrence of cristobalite with clinoptilolite in deep-sea drilling cores suggests that this zeolite may also be a silica buffer. In carbonate host rocks where quartz nuclei are abundant and where silica buffers are absent, the silica concentration in pore fluid could have been reduced to the quartz solubility value (of about 14 ppm), so that only quartz, not cristobalite, would precipitate as chert. An alternative hypothesis relates chert mineralogy to the pH of interstitial fluids. Dickson (1954) performed laboratory experiments on silica precipitation from supersaturated solutions. He found that cristobalite readily crystallized as a metastable phase from alkaline solutions, whereas quartz precipitated as the stable phase in acidic solutions. Is it possible that the interstitial fluids in shaly and carbonate host rocks have very different pH values? A test of the various hypotheses depends upon the silica analysis of the pore fluids from JOIDES cores.

Returning finally to the Maiolica chert, we could apply the idea of a local concentration gradient around quartz nuclei as a working hypothesis. Turbidite and slump beds are common in the Maiolica facies (Bernoulli, 1972). Chert could have nucleated around clastic quartz debris. Indeed I have found, in numerous instances, a core of radiolarite, or of older chert, rimmed by an ellipsoidal shell of chert (Fig. 11). This radiolarite and older chert were debris carried down by slumping or by turbidity currents into the Maiolica basin of

Figure 11. Growth of chert nodules in Maiolica Limestone. A. Ellipsoidal nodule with siliceous limestone core. Silica has apparently been removed from light-colored area to form chert rind around siliceous limestone nucleus. B. Ellipsoidal nodule with radiolarite core. Radiolarite fragment (center) in turbidite bed serves as nucleus for chert (rim) growth in Maiolica Limestone. C. Ellipsoidal nodule with core of deformed chert. Fragment of early diagenetic chert (center) disrupted by submarine sliding and present as clast in turbidite bed served as nucleus for chert (rim) growth in Maiolica Limestone.

pelagic ooze sedimentation. They were probably composed of chalcedonic quartz and served as nuclei, while the biogenic debris in the enclosing matrix supplied the silica, for chert growth.

I had thought that the abundance of chert in the Alpine Lower Cretaceous formations might be related to the Jurassic silica crisis. For example, I intuitively thought that a depletion of silica in ocean water would lead to a steeper concentration gradient in pore water near the sediment-water interface and thus facilitate the transport of silica for chertification near the sea floor. However, quantitative computations showed that the effect would be insignificant, and the chertification of the Maiolica facies could not be demonstrably related to the silica budget of the Early Cretaceous Tethys.

Conclusion

In a speculative mood, I have visualized an ammonitico rosso sea somewhat restricted, like the present eastern Mediterranean with its home-made bottom currents. I ventured a risky guess that the ophiolite ocean was 2,600 m deep in its central trough. I imagined a sudden invasion of equatorial currents into this Late Jurassic Tethys. I fancied a population explosion of the siliceous flora and a silica crisis. I pictured the aggressive coccoliths usurping the throne of a declining Radiolaria dynasty in the Tethys. I made a few tentative computations and found to my surprise that my intuition was wrong: chertification of the Maiolica facies did not seem to have much to do with the Tithonian silica crisis; rather, the nodular chert growth might have been triggered by the introduction of quartz-bearing debris into an environment in which amorphous silica occurred biogenically. I have written an essay, not a report on a specific research project. There are few new facts, but some new points of view and some new methods of attack. My interpretations can be wrong in a small or large way. However, the main conclusion I have come to is clear: we can learn a lot about Alpine history and about pelagic sedimentation through a study of the Alpine pelagic rocks, and the key lies in paleoceanography.

Acknowledgments

I am greatly indebted to all the participants of the 1973 Zurich Symposium on Pelagic Sediments; their written contributions (in Hsü and Jenkyns, 1974) and oral discussions were my sources of inspiration. I thank Jack Hayes for his insistence that I speak on the theme of Alpine pelagic rocks during the 1974 Society of Economic Paleontologists and Mineralogists symposium at San Antonio; his command was the motivation. Hugh Jenkyns, Dan Bernoulli, Rudolf Trümpy, Bob Garrison, Ueli Briegel, Maria Cita, Bill Ryan, and several others read a first draft of the manuscript and made valuable suggestions. Finally, I thank R. Gygi, K. Kelts, and A. Lambert for several of the illustrations used in this article.

References Cited

Aubouin, J., 1964, Réflexions sur la faciés "ammonitico rosso": Soc. Géol. France Bull. v. 6, p. 475–501.

Aubouin, J., Brunn, J.-H., Celet, P., Dercourt, J., Godfriaux, I., and Mercier, J., 1963, Esquisse de la géologie de la Grèce, *in* Durand Delga, M., ed., Paul Fallt memorial volume (Pt. 2): Géol. Soc. France, p. 583–610.

Bader, R. G., and others, 1970, Initial reports of the Deep Sea Drilling Project, Vol. 4: Washington, D.C., U.S. Govt. Printing Office, 753 p.

Beall, A. O., and Fischer, A. G., 1969, Sedimentology, *in* Ewing, M., and others, Initial reports of the Deep Sea Drilling Project, Vol. 1: Washington, D.C., U.S. Govt. Printing Office, p. 521–593.

Berger, W. H., 1967, Foraminiferal ooze: Solution at depth: Science, v. 156, p. 383–385.

____1970a, Biogenous deep-sea sediments: Fractionation by deep-sea circulation: Geol. Soc. America Bull., v. 81, p. 1385–1402.

____1970b, Planktonic Foraminifera: Selective solution and the lysocline: Marine Geology, v. 8, p. 111–138.

Berger, W. H., and von Rad, U., 1972, Cretaceous and Cenozoic sediments from the Atlantic Ocean, *in* Hayes, D. E., Pimm, A. C., and others, Initial reports of the Deep Sea Drilling Project, Vol. 14: Washington, D.C., U.S. Govt. Printing Office, p. 787–954.

Bernoulli, D., 1972, North Atlantic and Mediterranean Mesozoic facies: A comparison, *in* Hollister, C. D., Ewing, J. I., and others, Initial reports of the Deep Sea Drilling Project, Vol. 11: Washington, D.C., U.S. Govt. Printing Office, p. 801–822.

Bernoulli, D., and Jenkyns, H. C., 1974, Alpine Mediterranean and central Atlantic Mesozoic facies in relation to early evolution of Tethys: Soc. Econ. Paleontologists and Mineralogists Spec. Pub. 19, p. 129–160.

Bischoff, J. L., and Ku, T.-L., 1970, Pore fluids of Recent marine sediments; I, Oxidizing sediments of 20° N, continental rise to Mid-Atlantic Ridge: Jour. Sed. Petrology, v. 40, p. 960–972.

Bischoff, J. L., and Sayles, F. I., 1972, Pore fluid and mineralogical studies of Recent marine sediments: Baucr depression region of East Pacific Rise: Jour. Sed. Petrology, v. 42, no. 3, p. 711–724.

Boström, K., 1967, The problem of excess manganese in pelagic sediments, *in* Abelson, P. H., ed., Researches in geochemistry (Vol. 2): New York, John Wiley & Sons, p. 421–452.

____1973, The origin and fate of ferromanganoans at active ridge sediments: Stockholm Contr. Geology, v. 27, p. 149–243.

Boström, K., Peterson, M.N.A., Joensuu, O., and Fischer, D. E., 1969, Aluminum-poor ferromanganoan sediments on active oceanic ridges: Jour. Geophys. Research, v. 74, p. 3261–3270.

Bukry, D., Douglas, R. G., Kling, S. A., and Krasheninnikov, V., 1970, Planktonic microfossil biostratigraphy of the northwestern Pacific Ocean, *in* Fischer, A. G., and others, Initial reports of the Deep Sea Drilling Project, Vol. 6: Washington, D.C., U.S. Govt. Printing Office, p. 1253–1300.

Cadisch, J., Eugster, H., and Wenk, E., 1968, Geologischer Atlas der Schweiz, 1:25,000, Blatt Scuol-Schuls-Tarasp (Atlasblatt 44), Erläuterungen: Bern, Schweiz. Geol. Komm., 66 p.

Calvert, S. E., 1974, Deposition and diagenesis of silica in marine sediments, *in* Hsü, K. J. and

Jenkyns, H. C., Pelagic sediments: On land and under the sea: Internat. Assoc. Sedimentologists Spec. Pub. 1, p. 273–300.

Chen, C., 1964, Pteropod oozes from Bermuda pedestal: Science, v. 144, p. 60–62.

Colom, G., 1958, Sur les caractères de la sédimentation des géosynclinaux mésozoïques: Soc. Géol. France Bull., ser. b, v. 7, p. 1167–1187.

Cronan, D. S., 1973, Basal ferruginous sediments cored during Leg 16, *in* van Andel, T. H., Heath, G. R., and others, Initial reports of the Deep Sea Drilling Project, Vol. 16: Washington, D.C., U.S. Govt. Printing Office, p. 601–604.

Davis, E. F., 1918, The radiolarian cherts of the Franciscan group: California Univ. Dept. Geol. Sci. Bull., v. 11, p. 235–432.

Dewey, J. F., 1969, Evolution of the Appalachian/Caledonian orogen: Nature, v. 222, p. 124–129.

Dewey, J. F., Pitman, W. C., III, Ryan, W.B.F., and Bonnin, J., 1973, Plate tectonics and the evolution of the Alpine System: Geol. Soc. America Bull., v. 84, p. 3137–3180.

Dickson, F. W., 1954, Geochemical and petrographical aspects of mercury ore deposits [Ph.D. thesis]: Los Angeles, California Univ., 174 p.

Dumont, J. F., Gutnic, M., Marcoux, J., Monod, O., and Poisson, A., 1974, Le Trias des Taurides occidentales (Turquie). Définition du bassin pamphylien: Un nouveau domaine à ophiolithes à la marge externe de la chaîne taurique: Deutsch. Geol. Gesell. Zeitschr., v. 123, p. 385–409.

Dunham, R. J., 1969, Early vadose silt in Towsend Mound (reef), New Mexico, *in* Friedman, G. M., ed., Depositional environments in carbonate rocks: Soc. Econ. Paleontologists and Mineralogists Spec. Pub. 14, p. 139–181.

Fanning, K. A., and Pilson, M. E., 1971, Interstitial silica and pH in marine sediments: Some effects of sampling procedures: Science, v. 173, p. 1228–1230.

Fischer, A. G., 1963, The Lofer cyclothems of the Alpine Triassic: Kansas Geol. Survey Bull. 169, p. 107–149.

Fischer, A. G., and Garrison, R. E., 1967, Carbonate lithification on the sea floor: Jour. Geology, v. 74, p. 488–496.

Fischer, A. G., and others, 1971, Initial reports of the Deep Sea Drilling Project, Vol. 6: Washington, D.C., U.S. Govt. Printing Office, 1329 p.

Francheteau, J., 1970, Paleomagnetism and plate tectonics: Marine Phys. Lab., Scripps Inst. Oceanography, Ref. 70–30, 345 p.

Fuchs, T., 1877, Ueber die Entstehung der Aptychenkalk: Sitzungsbericht Akad. Wiss. Wien, Math.-Naturw. Kl., Abt. 1, v. 76, p. 329–334.

——1882, Welche Ablagerungen haben wir als Tiefseebildungen zu betrachten?: Neues Jahrb. Mineralogie, Geologie u. Paläontologie, v. II, Beilage Band, p. 487–784.

——1894, Ueber einige von der oesterreichischen Tiefsee Expedition S.M. Schiffes "Pola" in bedeutenden Tiefen gedredschten Cylindrites-aehnliche Körper: Berichte Komm. Erforschung oestlichen Mittelmeers. Denkschrift Akad. Wiss. Wien, Math.-Naturw. Kl., v. 61, p. 11–22.

——1898, Einige Bemerkungen über das Projekt eines internationalen schwimmenden Laboratoriums: Separat-Abdruck Verh. k.k. Geol. Reichanstalt, 1898, nr. 4., 8 p. (reprint).

Garrison, R. E., 1971, Cherts and radiolarian mudstones, *in* Fischer, A. G., and others, Initial reports of the Deep Sea Drilling Project, Vol. 6: Washington, D.C., U.S. Govt. Printing Office, p. 1212–1218.

Garrison, R. E., and Fischer, A. G., 1969, Deep-water limestones and radiolarites of the Alpine Jurassic. Soc. Econ Paleontologists and Mineralogists Spec. Pub. 14, p. 20.

Gervitz, J. L., and Friedman, G. M., 1966, Deep-sea carbonate sediments of the Red Sea and their implications on marine lithification: Jour. Sed. Petrology, v. 36, no. 1, p. 143–151.

Gibson, T. G., and Towe, K. M., 1971, Eocene volcanism and the origin of a Horizon A: Science, v. 172, p. 152–154.

Goldberg, E. D., and Arrhenius, G.O.S., 1958, Chemistry of Pacific pelagic sediments: Geochim. et Cosmochim. Acta, v. 13, p. 153–212.

Grabau, A. W., and O'Connell, M., 1917, Were the graptolitic shales, as a rule, deep- or shallow-water deposits?: Geol. Soc. America Bull., v. 28, p. 959–964.

Grunau, H. R., 1947, Geologie von Arosa (Graubünden) mit besonderer Berücksichtigung des Radiolarit-Problems: Zürich, Aschmann und Scheller, 108 p.

____1965, Radiolarian cherts and associated rocks in space and time: Eclogae Geol. Helvetiae, v. 58, no. 1, p. 157–208.

Hallam, A., 1967, Sedimentology and paleogeographic significance of certain red limestones and associated beds in the Lias of the Alpine region: Scottish Jour. Geology, no. 3, p. 195–220.

Heath, G. R., 1973, Cherts from the eastern Pacific, Leg 16, Deep Sea Drilling Project, *in* van Andel, T. H., Heath, G. R., and others, Initial reports of the Deep Sea Drilling Project, Vol. 16: Washington, D.C., U.S. Govt. Printing Office, p. 609–613.

____1974, Dissolved silica and deep-sea sediments, *in* Hay, W. W., ed., Studies in paleo-oceanography: Soc. Econ. Paleontologists and Mineralogists Spec. Pub. 20, p. 77–93.

Heath, G. R., and Moberly, R., Jr., 1971, Cherts from the western Pacific, Leg 7, Deep Sea Drilling Project, *in* Winterer, E. L., and others, Initial reports of the Deep Sea Drilling Project, Vol. 7: Washington, D.C., U.S. Govt. Printing Office, 1757 p.

Heezen, B. C., MacGregor, I. D., Foreman, H. P., Forristal, G., Hekel, H., Hoskins, R. H., Jones, E.J.W., Kaneps, A., Krasheninnikov, V. A., Okada, H., and Ruef, M. H., 1973, Diachronous deposits; A kinematic interpretation of the post Jurassic sedimentary sequence on the Pacific plate: Nature, v. 241, p. 25–32.

Hollister, C. D., Ewing, J. I., and others, 1972, Initial reports of the Deep Sea Drilling Project, Vol. 11: Washington, D.C., U.S. Govt. Printing Office, 1077 p.

Hollman, R., 1962, Ueber Subsolution und die "Knollenkalk" des calcare Ammonitico Rosso Superiore im Monte Baldo (Malm, Norditalien): Neues Jahrb. Geologie u. Paläontologie Monatsh., v. 4, p. 163–179.

____1964, Subsolutions-Fragmente (Zur Biostratonomie der Ammonoiden im Malm des Monte Baldo-Norditalien) Neues Jahrb. Geologie u. Paläontologie Abh., v. 119, p. 22–82.

Hsü, K. J., 1971, Origin of the Alps and the western Mediterranean: Nature, v. 233, p. 44–48.

Hsü, K. J., and Andrews, J. E., 1970, Lithology of the Mid-Atlantic Ridge sequence, *in* Maxwell, A. E., and others, Initial reports of the Deep Sea Drilling Project, Vol. 3: Washington, D.C., U.S. Govt. Printing Office, p. 445–453.

Hsü, K. J., and Jenkyns, H. C., 1974, Introduction, *in* Hsü, K. J., and Jenkyns, H. C., Pelagic sediments: On land and under the sea: Internat. Assoc. Sedimentologists Spec. Pub. 1, p. 1–10.

Jeffreys, J. G., 1877, Presidential address to the Biological Section: British Assoc. Adv. Sci. Rept., p. 79.

Jenkyns, H. C., 1971, The genesis of condensed sequences in the Tethyan Jurassic: Lethaia, v. 4, p. 327–352.

____1974, Origin of red nodular limestones (ammonitico rosso, Knollenkalk) in the Mediterranean Jurassic, a diagenetic model, *in* Hsü, K. J., and Jenkyns, H. C., Pelagic sediments: On land and under the sea: Internat. Assoc. Sedimentologists Spec. Pub. 1, p. 249–272.

Knopf, A., 1948, The geosynclinal theory: Geol. Soc. America Bull., v. 59, p. 649–670.

Kossmat, F., 1921, Die mediterranen Kettengebirge in ihrer Beziehung zum Gleichgewichtszustand der Erdrinde: Sächsische Akad. Wiss., Math.-Phys. Kl. 38, no. 2, p. 46–48.

Krauskopf, K. B., 1956, Dissolution and precipitation of silica at low temperatures: Geochim. et Cosmochim. Acta, v. 10, p. 1–26.

Lacombe, H., and Tchernia, P., 1973, Caractères hydrologiques et circulation des eax en Méditerranée, *in* Stanley, D. J., ed., The Mediterranean Sea: Stroudsburg, Pa., Dowden, Hutchinson, and Ross, p. 25–36.

Lancelot, Y., 1973, Chert and silica diagenesis in sediments from the central Pacific, *in* Winterer, E. L., Ewing, J. I., and others, Initial Reports of the Deep Sea Drilling Project, Vol. 17: Washington, D.C., U.S. Govt. Printing Office, p. 377–405.

Lancelot, Y., Hathaway, J. C., and Hollister, C. D., 1972, Lithology of sediments from the western North Atlantic, *in* Hollister, C. D., Ewing, J. I., and others, Initial reports of the Deep Sea Drilling Project, Vol. 11: Washington, D.C., U.S. Govt. Printing Office, p. 901–949.

Lisitsin, A. P., 1971, Distribution of siliceous microfossils in suspension and in bottom sediments, *in* Funnell, B. M., and Riedel, W. R., eds., Micropaleontology of oceans: Cambridge, Cambridge Univ. Press, p. 173–195.

Livingston, D. A., 1963, Chemical composition of rivers and lakes: U.S. Geol. Survey Prof. Paper 440-G, 64 p.

Lucas, G., 1966, Fonds durcis, lacunes sous-marines, séries condensées et ondes marines ou "Tsunami": Acad. Sci. Comptes Rendus, v. 262, p. 2141–2144.

Maxwell, A. E., and others, 1970, Initial reports of the Deep Sea Drilling Project: Dakar, Senegal, to Rio de Janeiro, Brazil, Vol. 3: Washington, D.C., U.S. Govt. Printing Office, 806 p.

Miller, A. R., 1973, Speculations concerning bottom circulation in the Mediterranean Sea, *in* Stanley, D. J., ed., The Mediterranean Sea: Stroudsburg, Pa., Dowden, Hutchinson, and Ross, p. 37–42.

Milliman, J. D., and Müller, J., 1973, Precipitation and lithification of the eastern Mediterranean Sea: Sedimentology, v. 20, p. 29–45.

Moores, E. M., 1973, Plate tectonics significance of Alpine peridotite types: Earth Sci. Rev., v. 9, p. 963–975.

Müller, J., and Fabricius, F., 1974, Magnesian-calcite nodules in the Ionian deep sea: An actualistic model for the formation of some nodular limestones, *in* Hsü, K. J., and Jenkyns, H. C., Pelagic sediments: On land and under the sea: Internat. Assoc. Sedimentologists Spec. Pub. 1, p. 235–247.

Murray, J., and Renard, A. F., 1891, Report on deep-sea deposits based on the specimens collected during the voyage of H.M.S. *Challenger* in the years 1872–1876: London, Eyre and Spottiswoods, 525 p.

Natterer, K., 1899, Chemisch-Geologische Tiefsee-Forschung: Expeditionen der Schiffe "Pola" und "Taurus" in das östliche Mittelmeer, Mamarameer und Rote Meer. Sonder-Abdruck aus der "Geographischen Zeitschrift." V. Jahrgang, 4 und 5, Heft: Leipzig, B. G. Teubner, p. 190–218.

Neumayr, M., 1873, Die Fauna der Schichten mit Aspidoceras acanthicum: Abh. k.k. geol. Reichanstalt, p. 141–257.

——1886, Erdgeschichte I. Band Allgemeine Geologie: Leipzig, Bibliographisches Institut, 544 p.

Nisbet, E. G., and Price, I., 1974, Siliceous turbidites: Bedded cherts as redeposited ocean ridge–derived sediments, *in* Hsü, K. J., and Jenkyns, H. C., Pelagic sediments: On land and under the sea: Internat. Assoc. Sedimentologists Spec. Pub. 1, p. 351–366.

Olausson, E., 1961, Studies of deep-sea cores: Swedish Deep-Sea Exped., 1947–48, Rept., v. 8, p. 335–391.

Pessagno, E. A., 1973, Age and geologic significance of radiolarian cherts in the California Coast Ranges: Geology, v. 1, p. 153–156.

Peterson, M.N.A., 1966, Calcite: Rates of dissolution in a vertical profile in the central Pacific: Science, v. 154, p. 1542–1544.

Peterson, M.N.A., and others, 1970, Initial reports of the Deep Sea Drilling Project, Vol. 2: Washington, D.C., U.S. Govt. Printing Office, 491 p.

Prevost, C., 1845, De la chronologie des terrains et du synchronisme des formations: Acad. Sci. Comptes Rendus, v. 20, Séance du avril 1845, p. 1–10.

Renz, O., 1972, Aptychi (ammonoids) from the Upper Jurassic and the Lower Cretaceous of the western North Atlantic, *in* Hollister, C. D., Ewing, J. I., and others, Initial reports of the Deep Sea Drilling Project, Vol. 11: Washington, D.C., U.S. Govt. Printing Office, p. 607–629.

Rex, R. W., 1970, X-ray mineralogy studies, *in* Peterson, M.N.A., and others, Initial reports of the Deep Sea Drilling Project, Vol. 2: Washington, D.C., U.S. Govt. Printing Office, p. 329–346.

Riedel, W. R., 1963, The preserved record: Paleontology of pelagic sediments, *in* Hill, M. N., ed., The sea (Vol. 3): New York, Interscience Pubs., Inc., p. 866–887.

Rose, G., 1858, Ueber die heteromorphen Zustände der kohlsäuren Kalkerde: Abh. königlichen Akad. Wiss., Berlin, p. 63–111.

Ruedemann, R., 1934, Paleozoic plankton of North America: Geol. Soc. America Mem. 2, p. 43–51.

Rüppell, E., 1829, Abbildungen und Beschreibung einiger neuer oder wenig gekannten Versteinerungen aus der Kalkschieferformation von Soleuhofen: Frankfurt am Main, 12 p.

Ryan, W.B.F., 1969, The floor of the Mediterranean Sea [Ph.D. thesis]: New York, Columbia Univ., 233 p.

Ryan, W.B.F., Hsü, K. J., and others, 1973, Initial reports of the Deep Sea Drilling Project, Vol. 13: Washington, D.C., U.S. Govt. Printing Office, 1447 p.

Sapunov, I., and Ziegler, B., 1975, Stratigraphische Probleme in Ober-Jura des westlichen Balkangebirges: Geol. Rundschau (in press).

Scheumann, K. H., 1932, Ueber die petrogenetische Abteilung des roten Erzgebirgsgneisses: Mineralog. u. Petrog. Mitt., v. 42, p. 423–426.

Schlager, W., 1974, Preservation of cephalopod skeletons and carbonate dissolution on ancient Tethyan sea floor, *in* Hsü, K. J., and Jenkyns, H. C., Pelagic sediments: On land and under the sea: Internat. Assoc. Sedimentologists Spec. Pub. 1, p. 49–70.

Sclater, J. G., and Detrick, R., 1973, Elevation of mid-ocean ridges and the basement ages of JOIDES deep sea drilling sites: Geol. Soc. America Bull., v. 84, p. 1547–1554.

Sclater, J. G., and Francheteau, J., 1970, The implication of terrestrial heat flow observations on current tectonic and geochemical models of the crust and the upper mantle of the Earth: Royal Astron. Soc. Geophys. Jour., v. 20, p. 509–542.

Sclater, J. G., Anderson, R. N., and Bell, M. L., 1971, Elevation of ridges and evolution of the central eastern Pacific: Jour. Geophys. Research, v. 76, p. 7888–7915.

Siever, R., 1957, The silica budget in the sedimentary cycle: Am. Mineralogist, v. 42, p. 821–841.

Smith, A. G., 1971, Alpine deformation and the oceanic areas of the Tethys, Mediterranean, and Atlantic: Geol. Soc. America Bull., v. 82, p. 2039–2070.

Steinmann, G., 1905, Die geologische Bedeutung der Tiefsee-absätze und der ophiolithischen Massengesteines: Berichte Naturf. Gesell. Freiburg, v. 16, p. 44–65.

____1913, Ueber Tiefenabsätze des Oberjura im Appenin: Allg. Versammlung Geol. Verh. Marburg, p. 472–475.

____1925, Gibt es Fossile Tiefseeablagerungen von erdgeschichtlichen Bedeutung?: Geol. Rundschau, v. 16, p. 435–468.

____1927, Die ophiolithischen Zonen in dem mediterranen Kettengebirge: Internat. Geol. Cong., 14th, Madrid 1926, Comptes Rendus, v. 2, p. 637–667.

Suess, E., 1875, Die Entstehung der Alpen: Vienna, W. Braumüller, 168 p.

____1909, Das Antlitz der Erde, Pt. 2 (3rd ed.): Leipzig, G. Freytag, 789 p.

Taliaferro, N. L., 1943, Franciscan-Knoxville problem: Am. Assoc. Petroleum Geologists Bull., v. 27, p. 109–219.

Tietz, E., 1897, Der VII Internationale Geologenkongress in Petersburg: Verh. k.k. Geol. Reichanstalt, p. 286–307.

Trümpy, R., 1960, Paleotectonic evolution of the central and western Alps: Geol. Soc. America Bull., v. 71, p. 843–908.

——1971, Stratigraphy in mountain belts: Geol. Soc. London Quart. Jour., v. 126, p. 293–318.

Twenhofel, W. H., 1939, Environments of origin of black shales: Am. Assoc. Petroleum Geologists Bull., v. 23, p. 1178–1198.

von der Borch, C. C., and Rex, R. W., 1970, Amorphous iron oxide precipitates in sediments cored during Leg 5, Deep Sea Drilling Project, *in* McManus, D. A., and others, Initial reports of the Deep Sea Drilling Project, Vol. 5: Washington, D.C., U.S. Govt. Printing Office, p. 541–544.

von der Borch, C. C., Nesteroff, W. D., and Galehouse, J. S., 1971, Iron rich sediments cored during Leg 8 of the Deep Sea Drilling Project, *in* Tracey, J. I., Jr., and others, Initial reports of the Deep Sea Drilling Project, Vol. 8: Washington, D.C., U.S. Govt. Printing Office, 1037 p.

Ulrich, E. D., 1911, Revision of the Paleozoic systems: Geol. Soc. America Bull., v. 22, p. 281–680.

Wähner, F., 1886, Zur heteropischen Differenzierung des alpinen Lias: Verh. k.k. Geol. Reichanstalt, p. 168–176, 190–206.

Walther, J., 1897, Ueber die Lebensweise fossiler Meerestiere: Zeitschr. Deutsch. Geol. Gesell., v. 49, p. 209–273.

Wise, S. W., and Hsü, K. J., 1971, Genesis and lithification of a deep sea chalk: Eclogae Geol. Helvetiae, v. 64, p. 273–278.

Wise, S. W., and Weaver, F. M., 1974, Chertification of oceanic sediments, *in* Hsü, K. J., and Jenkyns, H. C., Pelagic sediments: On land and under the sea: Internat. Assoc. Sedimentologists Spec. Pub. 1, p. 301–326.

Wollast, R., and Garrels, R. M., 1971, Diffusion coefficient of silica in sea water: Nature, v. 229, p. 94.

MANUSCRIPT RECEIVED BY THE SOCIETY MAY 21, 1974
REVISED MANUSCRIPT RECEIVED DECEMBER 3, 1974
MANUSCRIPT ACCEPTED MARCH 12, 1975